Microwave Imaging and Electromagnetic Inverse Scattering Problems

Microwave Imaging and Electromagnetic Inverse Scattering Problems

Special Issue Editors

Loreto Di Donato
Andrea F. Morabito

MDPI • Basel • Beijing • Wuhan • Barcelona • Belgrade

Special Issue Editors
Loreto Di Donato
University of Catania
Italy

Andrea F. Morabito
Università degli Studi Mediterranea di Reggio Calabria
Italy

Editorial Office
MDPI
St. Alban-Anlage 66
4052 Basel, Switzerland

This is a reprint of articles from the Special Issue published online in the open access journal *Journal of Imaging* (ISSN 2313-433X) in 2019 (available at: https://www.mdpi.com/journal/jimaging/special_issues/Microwave_Imaging).

For citation purposes, cite each article independently as indicated on the article page online and as indicated below:

LastName, A.A.; LastName, B.B.; LastName, C.C. Article Title. *Journal Name* **Year**, *Article Number*, Page Range.

ISBN 978-3-03921-950-6 (Pbk)
ISBN 978-3-03921-951-3 (PDF)

Contents

About the Special Issue Editors

Loreto Di Donato received his B.S. and M.S. Laurea degrees in Biomedical Engineering from the University of Naples "Federico II" (Italy) in 2006 and 2008, respectively. After a year spent at the Electromagnetic Diagnostic Research Group of the Institute for the Electromagnetic Sensing of the Environment—National Research Council (IREA-CNR) of Italy in Naples as a research assistant, he received his Ph.D. in Information Engineering from the University "Mediterranea" of Reggio Calabria in 2013. He has been with the Department of Electrical, Electronics and Computer Engineering (DIEEI) as Assistant Professor of Electromagnetic Fields at the University of Catania (Italy) since his appointment in 2013, and starting from 2021, he shall serve as Associate Professor here. His research activities are mainly focused on inverse problems in electromagnetics, with particular interest in inverse scattering problems and microwave imaging for biomedical, subsurface, and plasma diagnostics, as well as focusing and antenna synthesis problems. Loreto Di Donato was a Young Scientist Awarded at the XXX URSI General Assembly in 2011, received the prize "Gaetano Latmiral" from the Electromagnetic Italian Society in 2016 and obtained the Abilitazione Scientifica Nazionale (ASN) for the academic position Associate Professor in 2017.

Andrea F. Morabito received his Laurea degree in Telecommunications Engineering (summa cum laude) and Ph.D. degree in Computer, Biomedical, and Telecommunications Engineering from the University of Reggio Calabria, Italy, where he has been Assistant Professor in Electromagnetic Fields since 2010. His research work is mainly focused on models and effective strategies for electromagnetic forward and inverse scattering problems, as well as on antenna theory, design, optimal synthesis and therapeutic applications, with applications ranging from biomedical imaging to radar and satellite telecommunications. Dr. Morabito has been a visiting researcher at the Eledia Research Center of the University of Trento and is a member of the LEMMA Research Group of the Italian Electromagnetics Society, and of the European Association on Antennas and Propagation. He is author of about 100 papers published on peer-reviewed scientific journals or international conference proceedings. Dr. Morabito has been awarded by the Italian Electromagnetics Society with both the "Barzilai" and "Latmiral" Prizes, and received the Abilitazione Scientifica Nazionale (ASN) for the academic position Associate Professor.

Preface to "Microwave Imaging and Electromagnetic Inverse Scattering Problems"

Microwave imaging (MWI) is a promising technology able to perform noninvasive diagnostics in many disparate fields, such as nondestructive testing evaluation, civil and geophysical prospecting, surveillance and security monitoring, biomedical imaging, and many others. The main advantages of MWI lies in its noninvasiveness, low cost, as well as the on-field portability of the involved equipment. For this reason, over the last thirty years, a large effort has been spent in studying and understanding the ultimate achievable performance of MWI, and hence to develop several reconstruction strategies able to overcome the main drawbacks concerning ill-posedness, false solutions, resolution limits, and heavy computational burden.

Nowadays, a plethora of approaches for MWI co-exist in the scientific literature and an ever-increasing number of experimental tests and facilities, such as clinical trials, are under development. This notwithstanding, microwave imaging and inverse scattering problems still deserve great efforts from a methodological point of view, in order to develop effective and efficient solutions able to provide reconstructions as reliable, accurate, and fast as possible. Indeed, despite the intrinsic low resolution of microwave diffraction tomography, emerging signal processing techniques and innovative hardware solutions, such as, for example, compressive sensing and deep/machine learning, metamaterials and metasurfaces, millimeter and THz waves, as well as the employment of high-performing computers, may allow these huge drawbacks to be overcome, paving the way for the full advantages of MWI to be expressed in the above-cited fields of application.

This Special Issue presents ten papers by some of the most prominent international scientists and researchers in the area of microwave imaging and diagnostics. The topics include experimental, methodological, and emerging fields of application. The presented papers provide a glimpse into the current state of the art in microwave imaging and electromagnetic inverse scattering and the different nature of the involved contexts and applications.

Loreto Di Donato, Andrea F. Morabito
Special Issue Editors

Journal of
Imaging

MDPI

Article

A Tomograph Prototype for Quantitative Microwave Imaging: Preliminary Experimental Results

Alessandro Fedeli [1], Manuela Maffongelli [2], Ricardo Monleone [2], Claudio Pagnamenta [2], Matteo Pastorino [1,*], Samuel Poretti [2], Andrea Randazzo [1] and Andrea Salvadè [2]

[1] Department of Electrical, Electronic, Telecommunications Engineering, and Naval Architecture, University of Genoa, 16145 Genoa, Italy; alessandro.fedeli@unige.it (A.F.); andrea.randazzo@unige.it (A.R.)

[2] Department of Technology and Innovation, University of Applied Sciences of Southern Switzerland, 6928 Manno, Switzerland; manuela.maffongelli@supsi.ch (M.M.); ricardo.monleone@supsi.ch (R.M.); claudio.pagnamenta@supsi.ch (C.P.); samuel.poretti@supsi.ch (S.P.); andrea.salvade@supsi.ch (A.S.)

* Correspondence: matteo.pastorino@unige.it

Received: 1 October 2018; Accepted: 22 November 2018; Published: 26 November 2018

Abstract: A new prototype of a tomographic system for microwave imaging is presented in this paper. The target being tested is surrounded by an ad-hoc 3D-printed structure, which supports sixteen custom antenna elements. The transmission measurements between each pair of antennas are acquired through a vector network analyzer connected to a modular switching matrix. The collected data are inverted by a hybrid nonlinear procedure combining qualitative and quantitative reconstruction algorithms. Preliminary experimental results, showing the capabilities of the developed system, are reported.

Keywords: microwave imaging; tomography; inverse problems

1. Introduction

In the last years, there has been growing interest in the development of microwave imaging systems in various applicative fields, such as for civil, industrial, and biomedical applications [1–4]. In fact, microwaves have the potential ability to penetrate into dielectric materials, allowing for directly obtaining information about the internal dielectric properties of the samples under test (SUT). Despite the potential advantages, there are still some problems to be faced, which motivate the ever-expanding research activities on this topic. First, in many cases, the targets are composed of inhomogeneous, lossy, and dispersive media (e.g., when dealing with biological materials). The electromagnetic waves are thus typically subject to high attenuations during the propagation through the target, and the information about the material properties are contained in a complex way in the scattered field. Moreover, the dielectric contrast with respect to the external embedding may be high, requiring the adoption of matching layers. Consequently, it is necessary to properly design the antennas, the embedding structures, and the transmission/measurement hardware. Several experimental apparatuses have been proposed, based on both radar concepts and inverse scattering methods (see, for example, [5–7]). A recent overview containing several examples of experimental set ups has been reported in the literature [8], and the reader is referred to that book for a rather detailed description of these configurations.

Furthermore, appropriate inversion procedures, able to exploit the available a-priori information about the target, should be used in order to process the measurements made available by the hardware. In this framework, several interesting approaches have been recently introduced [9–18]. On the one hand, they have been designed in order to improve the capabilities of the so-called qualitative methods, which are aimed at retrieving only the presence/shape of targets, as well as some particular features of interest. For example, in non-destructive evaluations, the knowledge of the position and dimension of a defect or a crack could in fact represent sufficient information. On the other hand, much more sophisticated approaches have been devised for inspecting strong scatterers and producing images of the dielectric properties of the scene under test. This is of fundamental importance in biomedical applications, where microwave imaging techniques have been proposed for breast cancer [19–22] and brain stroke detection [23–26], as well as in several other diagnostic processes, such as the monitoring of microwave ablation [27], guiding endo-capsules [28], and so on.

In this paper, a new prototype of the microwave imaging system is presented. It allows for collecting frequency-stepped scattered-field data in a tomographic configuration by using a multi-static setup. The antennas have been specifically designed in order to minimize the reflections from the external surface, and to optimize the propagation inside the target. An ad-hoc 3D printed holding structure is used to keep the radiating elements in contact with the target. The measured data are inverted using a hybrid inverse-scattering procedure [29], which combines a delay-and-sum (DAS) qualitative algorithm [30] (providing a qualitative image of the dielectric discontinuities) with a quantitative method based on an inexact-Newton/Landweber (INLW) scheme [31] (providing a reconstruction of the distribution of the dielectric properties of the target). In particular, the obtained qualitative image allows for focusing the INLW method on the regions of the target in which the discontinuities with respect to the background are found. The capabilities of the developed system are assessed by means of preliminary experimental results.

The paper is organized as follows. The developed measurement hardware and the related inversion procedure are described in Section 2. Section 3 reports some of the experimental results aimed at validating both the system and the reconstruction algorithm. Finally, conclusions are drawn in Section 4.

2. Measurement Setup and Inversion Procedure

A block diagram of the developed prototype is shown in Figure 1. It is composed of a vector network analyzer (VNA), a radio frequency (RF) switch matrix connected to a set of ad-hoc antennas ($M = 16$ antennas are adopted in the current prototype), and a control board used to drive the switches. The whole system is managed by a personal computer (PC), which is used to select the active antenna pair (via universal serial bus (USB) connection), control the VNA, collect the measurements (via an Ethernet connection), and execute the reconstruction algorithms. The PC also takes care of the synchronization between the various components of the system.

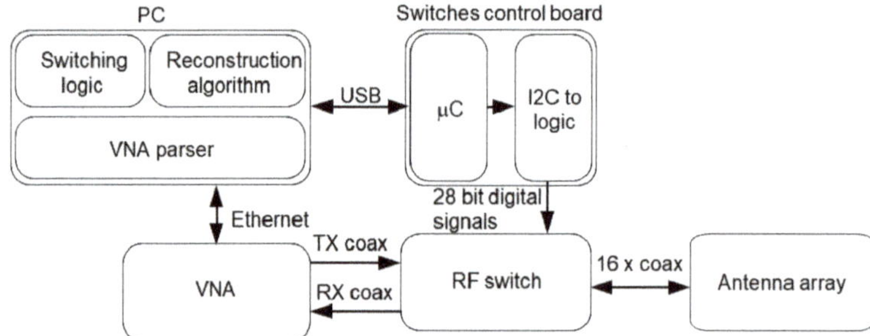

Figure 1. Block diagram of the microwave tomographic system architecture. USB—universal serial bus; µC—microcontroller; VNA—vector network analyzer; PC—personal computer; TX—transmission; RX—reception.

The RF switch matrix is composed of two levels, as shown in Figure 2. The first one, which is directly connected to the transmit and receive ports of the VNA, is composed of two 1:4 switch boards, whereas the second one is composed of four 2:4 switch boards, whose outputs are connected to the 16 antennas. For both of the levels, Peregrine Semiconductor's PE42441 SP4T absorptive RF switches have been used in the switch boards. The working frequency band of such components ranges from 10 MHz up to 8 GHz. Each board has been manufactured using a Rogers RO4350B substrate (characterized by complex relative dielectric permittivity $\epsilon_r = 3.66 - j0.011$ at 2.5 GHz) in order to reduce the RF signal attenuation and the changes in the electrical parameters. The switching subsystem allows for independently selecting a pair of antennas among the possible 240 combinations. One antenna at a time is used in transmit mode (i.e., it is connected to the transmitting (TX) port of the VNA), whereas the remaining ones are sequentially connected to the receiving port of the VNA for measuring the field scattered by the target. In order to reduce the interferences in the measurement, all of the inactive antennas are terminated using absorptive RF switch components, ensuring an isolation between the adjacent channels at the board connectors of about 90 dB at 3 GHz. The switch control board is equipped with an ATmega328 microcontroller programmed to transparently pass the USB data to an I2C bus, which is connected to four 16-bit PCF8575CDBE4 input/output (I/O) expanders used to drive the RF switches.

Folded quasi self-complementary antennas (FQSCA) [32] have been used in the prototype (whose layout is represented in Figure 3). Such antennas have been specifically developed in order to work in direct contact with the SUT, by reducing the reflections due to the mismatch with the target interface and optimizing the radiation inside the inspected structure. The operating bandwidth of the antennas ranges from 1.5 GHz up to 6 GHz. The radiating elements are kept in contact with the SUT using a custom 3D printed circular holding structure (shown in Figure 3) made of plastic (VeroWhite), with relative dielectric permittivity $\epsilon_r = 2.98$ and loss tangent $\tan\delta = 0.029$ at 2.4 GHz [33]. The holding structure has been produced using a stereolithographical process. The separation between the two adjacent antennas is 22.5° and the radius is 60 mm. In order to compensate for the different paths of the signal in the function of the selected antenna pairs, a calibration is performed. In particular, the calibration procedure is aimed at compensating the influence (i.e., phase shift and attenuation) of the switching electronics and cables for each pair of transmitting/receiving (TX/RX) antennas, thus resulting in measurements referenced to the antenna connectors. Actually, such a calibration is performed by subtracting from the amplitude (in decibel) and phase of the measured data (with and without inclusions) of each TX/RX pair the corresponding through measurements obtained by connecting together the RX and TX antenna connectors.

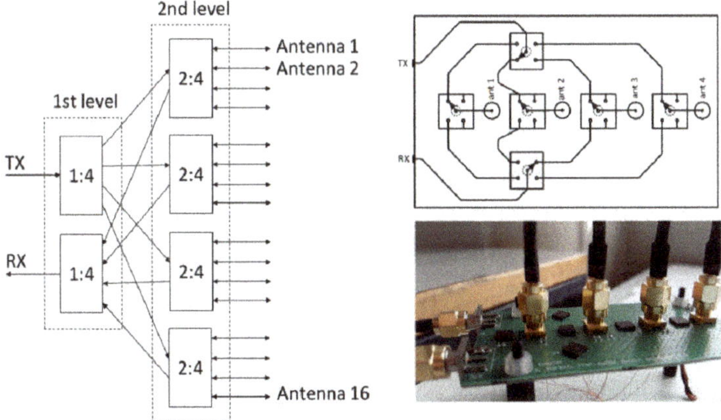

Figure 2. Block scheme of the switching board and photograph of the 2:4 switch module.

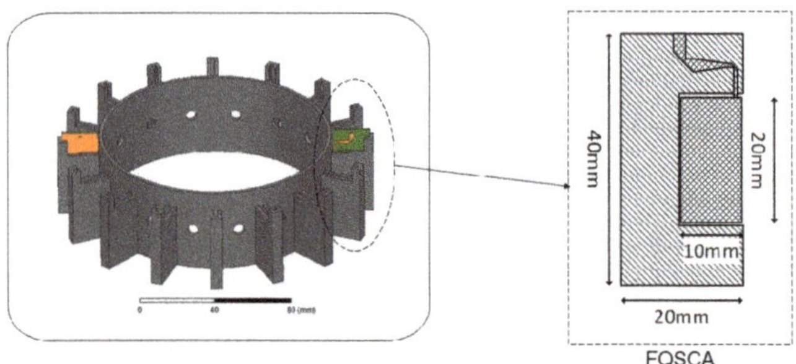

Figure 3. Schematic representation of the holding structure and layout of the folded quasi self-complementary antennas (FQSCA) antenna.

The scattered-field data (i.e., the difference between the measurements with and without the inclusions) collected by the system are processed using a hybrid qualitative–quantitative inversion scheme [29]. A flow chart of the developed inversion procedure is shown in Figure 4. Starting from the measured stepped-frequency data, an initial qualitative reconstruction $\Lambda(\mathbf{r})$, $\mathbf{r} \in \mathcal{D}$ ($\mathcal{D}$ being the inspected area), is obtained using a multistatic DAS method [30]. The time-domain response is synthesized by applying an inverse fast Fourier transform to the data. Such an image provides a rough indication of the eventually present inclusions.

Subsequently, a frequency-hopping (FH) inexact-Newton/Landweber method is used to obtain the quantitative distribution of the dielectric properties. In particular, for each considered frequency (hereafter, the superscript f denotes the frequency index), a hybrid INLW method is used to invert the non-linear and ill-posed scattering equation relating the unknown contrast function $\chi^f = \epsilon_r^f / \epsilon_{r,b}^f - 1$ (ϵ_r^f and $\epsilon_{r,b}^f$ being the complex relative dielectric permittivity of SUT and background, respectively) with the z-component of the scattered electric field e_s^f, as follows:

$$e_s^f(\mathbf{r}) = G_d^f \chi^f \left(I - G_s^f \chi^f \right)^{-1} e_b^f(\mathbf{r}) = F^f \left(\chi^f \right)(\mathbf{r}), \tag{1}$$

where e_b^f is the z-component of the electric field due to the background and $G_{d/s}^f$ are linear integral operators whose kernel is the Green's function for the background. In such an inversion procedure, the antennas have been modeled as ideal z-directed line-current sources. Consequently, the acquired data have been further pre-processed by properly scaling the measured values by a complex scaling factor, obtained by matching the data measured with the empty imaging chamber (filled by the background medium only) with the corresponding simulated electric field.

The INLW technique, outlined in the right side of Figure 4, is an iterative nonlinear inversion method formed by two nested loops. The outer one is based on a Newton scheme, which performs an iterative linearization of Equation (1) around the currently reconstructed value of the contrast function χ_i^f (i being the outer iteration index). The resulting linearized equation $F_i^{f\prime} h_i(\mathbf{r}) = e_s^f(\mathbf{r}) - F^f \left(\chi_i^f \right)(\mathbf{r})$, F_i^f being the Fréchet derivative of the scattering operator F^f in Equation (1), is solved in a regularized sense, using a truncated Landweber algorithm [31,34]. In particular, the regularized solution $h_i(\mathbf{r})$ is obtained by means of the following update formula (initialized with $h_{i,0} = 0$):

$$h_{i,l+1}(\mathbf{r}) = h_{i,l}(\mathbf{r}) - \beta_i F_i^{f*} \left(F_i^f h_{i,l} - e_s^f(\mathbf{r}) + F^f \left(\chi_i^f \right)(\mathbf{r}) \right), \; l = 0, 1, \ldots, N_{LW} \tag{2}$$

where F_i^{f*} is the adjoint of the operator F_i^f, $\beta_i = \| F_i^f \|^{-2}$ is a fixed relaxation coefficient, and N_{LW} is the maximum number of allowed iterations. In order to exploit the information obtained in the qualitative step, the Newton update has been modified by multiplying the solution of the linearized problem by the value of the normalized qualitative map Λ in the same point of the investigation domain (i.e., $\chi^f_{i+1}(\mathbf{r}) = \chi^f_i(\mathbf{r}) + \Lambda(\mathbf{r})h_i(\mathbf{r})$). This is the key aspect of the combination between the qualitative DAS method and the INLW algorithm. In fact, since $\Lambda(\mathbf{r}) \in [0,1]$, $\forall \mathbf{r} \in \mathcal{D}$, the contrast function χ is updated "faster" in the regions, in which the qualitative method found relevant discontinuities with respect to the background (i.e., when Λ is close to 1), and "slower" when Λ assumes low values, which means in the points outside the qualitatively-detected targets. Newton iterations (i.e., linearization, regularized solution, and update) then continue until a convergence criterion is met or a maximum number of iterations, N_{IN}, is reached. After that, the FH procedure requires that the obtained contrast function is used as the starting guess for the inversion at the subsequent frequency, until all of the available frequencies have been processed.

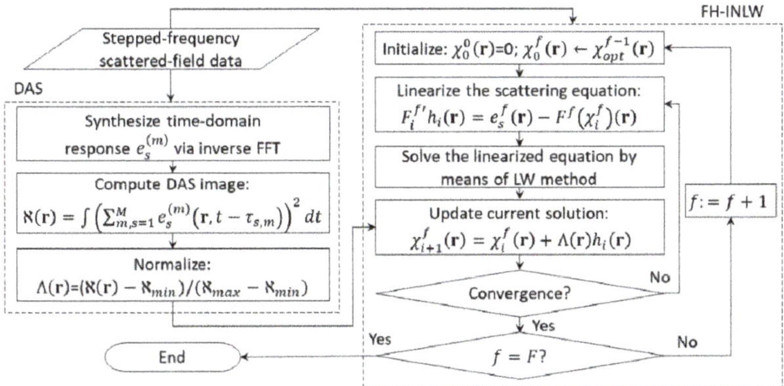

Figure 4. Flow chart of the inversion algorithm. DAS—delay-and-sum; FH-INLW—frequency-hopping inexact-Newton/Landweber.

3. Preliminary Experimental Results

The proposed system has been preliminarily tested with a SUT composed of a Plexiglas circular cylinder, with a diameter of $d = 120$ mm (Figure 5) filled with a mixture of oil, water, salt, and soy lecithin (relative dielectric permittivity $\epsilon_{r,b} \approx 3 - j0.73$ at 2.45 GHz). The liquid mixture, which occupies a vertical depth of 90 mm, has been prepared by adding small quantities of the single components to the mixture step-by-step, and checking the actual properties with a VNA dielectric probe (Keysight N1501A). In particular, a strong percentage of vegetable oil has been used, whereas other components, in smaller quantities, have been added to adjust the values to the desired ones. Salt has also been added in order to achieve the correct conductivity. Moreover, soy lecithin has been used as a surfactant for obtaining a stable mixture. The investigation domain, $\mathcal{D}$, is a circular region of the diameter, d, discretized into 1240 square subdomains with a 3.12 mm side. Three plastic tubes with a diameter of $d_i = 5$ mm, filled with a different liquid mixture (relative dielectric permittivity $\epsilon_r \approx 35 - j10$ in the considered frequency band), are approximately located at positions $\mathbf{r}_1 = (-20, -25)$ mm, $\mathbf{r}_2 = (20, -20)$ mm, and $\mathbf{r}_3 = (25, 20)$ mm. In particular, a 50%/50% mixture of vegetable oil and tap water with 2 g of soy lecithin has been used. A negligible dispersive behavior of the measured dielectric properties of both of the liquid mixtures has been observed in the considered frequency range. Therefore, dispersion has not been considered in the inversion procedure.

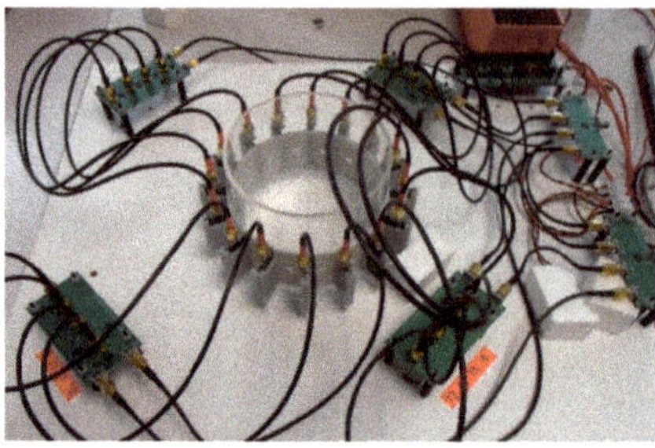

Figure 5. Photograph of the experimental set up.

The transmission scattering parameters have been acquired for all of the possible antenna pairs by considering 585 frequency steps in the range between 1 and 5 GHz. The acquisition time is mainly affected by the VNA measurement sweep time; the switch setting and I/O expander latency can be neglected. For the presented results, a sweep time of 1 s has been set for each of the antennas' combinations, resulting in a total measurement time of 240 s.

The measured values of S_{21} with and without the considered inclusions are shown in Figure 6 for a pair of opposite antennas (antennas #6 and #14). It is worth noting that most of the scattering contributions are located between 2 and 4 GHz. Consequently, the FH-INLW method has been applied, by considering the $F = 13$ equally-spaced frequencies in this range. The maximum number of inner and outer iterations in each INLW inversion step are $N_{LW} = 100$ and $N_{IN} = 100$, and the loops are stopped when the variation in the residual falls below 1%. The values of such parameters have been empirically selected on the basis of previous analyses. The results provided by the developed procedure are shown in Figures 7 and 8. In particular, the obtained qualitative map, showing the presence of the three inclusions, is reported in Figure 7. The quantitative reconstructions of the relative dielectric permittivity at the final frequency step are shown in Figure 8a (real part) and Figure 8b (imaginary part). As can be seen, the targets are accurately localized and the dielectric properties are quite correctly estimated. The root mean squared error (RMSE) on the reconstruction of the complex dielectric permittivity is equal to 0.48 in this case. The proposed hybrid inversion algorithm has also been compared with a bare INLW technique, in which the update of the solution in the Newton iterations is not weighted by the qualitative map [31]. The reconstructed distributions of the dielectric properties of the target are shown in Figure 9, and the corresponding RMSE is 0.52. This comparison evidences the effectiveness of the hybridization with the qualitative scheme, which allows for a significant improvement in both the estimation of the complex dielectric permittivity (especially, the real part) and the size of the target, as well as the reconstruction of the background medium.

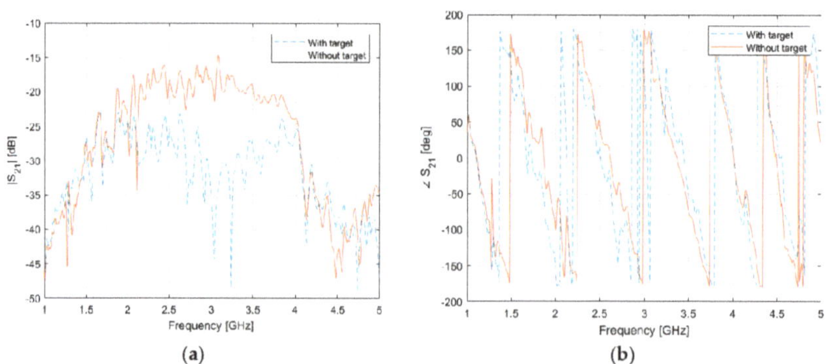

Figure 6. (**a**) Magnitude and (**b**) phase of the measured S_{21} for a couple of facing antennas with and without the targets.

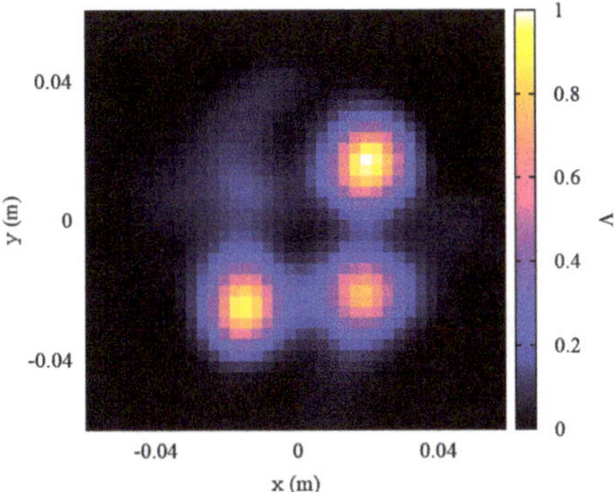

Figure 7. Experimental results. Qualitative indicator map.

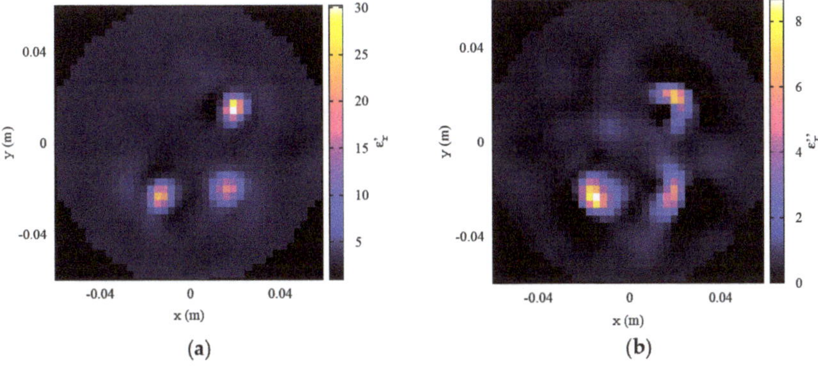

Figure 8. Experimental results. Reconstructed distributions of the (**a**) real part and (**b**) imaginary part of the complex relative dielectric permittivity at 3.8 GHz obtained with the proposed hybrid algorithm.

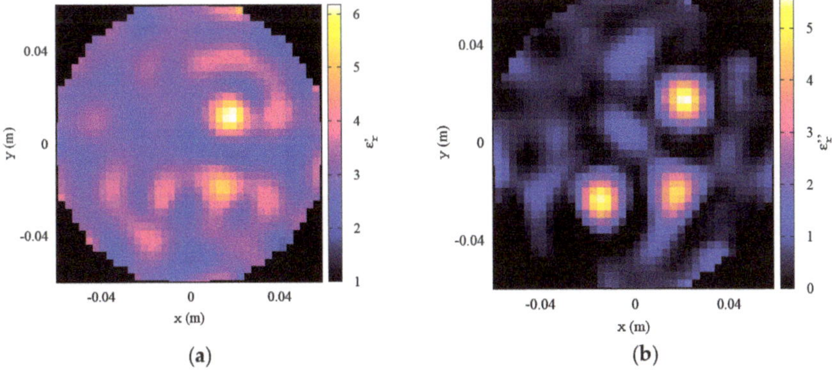

Figure 9. Experimental results. Reconstructed distributions of the (**a**) real part and (**b**) imaginary part of the complex relative dielectric permittivity at 3.8 GHz obtained with a bare quantitative inversion procedure.

4. Conclusions

A microwave system prototype for quantitative tomographic imaging has been presented in this paper. The system is composed of a set of ad-hoc antennas held in contact with the SUT by a custom 3D-printed structure, a modular switch matrix, a VNA, and a computer-based control and processing section. A hybrid qualitative/quantitative inversion method has been developed for processing the measured scattered-field data. The effectiveness of the imaging setup has been assessed by means of preliminary experimental results. Future developments will be aimed at performing a wider experimental validation campaign with more complex targets.

Author Contributions: Conceptualization, A.F., M.M., R.M., C.P., M.P., S.P., A.R., and A.S.; formal analysis, A.F., M.R., and A.R.; investigation, M.M., R.M., C.P., S.P., and A.S.; resources, M.M., R.M., C.P., S.P., and A.S.; software, A.F., M.R., and A.R.

Funding: This research received no external funding.

Conflicts of Interest: The authors declare no conflict of interest.

References

1. Meaney, P.M.; Goodwin, D.; Golnabi, A.H.; Zhou, T.; Pallone, M.; Geimer, S.D.; Burke, G.; Paulsen, K.D. Clinical Microwave Tomographic Imaging of the Calcaneus: A First-in-Human Case Study of Two Subjects. *IEEE Trans. Biomed. Eng.* **2012**, *59*, 3304–3313. [CrossRef] [PubMed]
2. Ghasr, M.T.; Horst, M.J.; Dvorsky, M.R.; Zoughi, R. Wideband microwave camera for real-time 3-D imaging. *IEEE Trans. Antennas Propag.* **2017**, *65*, 258–268. [CrossRef]
3. Gilmore, C.; Zakaria, A.; Pistorius, S.; LoVetri, J. Microwave Imaging of Human Forearms: Pilot Study and Image Enhancement. *Int. J. Biomed. Imaging* **2013**, *2013*, 673027. [CrossRef] [PubMed]
4. Benedetto, A.; Pajewski, L. *Civil Engineering Applications of Ground Penetrating Radar*; Springer: Cham, Switzerland, 2015; ISBN 978-3-319-04813-0.
5. Castorina, G.; Di Donato, L.; Morabito, A.F.; Isernia, T.; Sorbello, G. Analysis and design of a concrete embedded antenna for wireless monitoring applications. *IEEE Antennas Propag. Mag.* **2016**, *58*, 76–93. [CrossRef]
6. Monleone, R.D.; Pastorino, M.; Fortuny-Guasch, J.; Salvade, A.; Bartesaghi, T.; Bozza, G.; Maffongelli, M.; Massimini, A.; Carbonetti, A.; Randazzo, A. Impact of background noise on dielectric reconstructions obtained by a prototype of microwave axial tomograph. *IEEE Trans. Instrum. Meas.* **2012**, *61*, 140–148. [CrossRef]
7. Fear, E.C.; Bourqui, J.; Curtis, C.; Mew, D.; Docktor, B.; Romano, C. Microwave breast imaging with a monostatic radar-based system: A study of application to patients. *IEEE Trans. Microw. Theory Tech.* **2013**, *61*, 2119–2128. [CrossRef]
8. Pastorino, M.; Randazzo, A. *Microwave Imaging Methods and Applications*; Artech House: Boston, MA, USA, 2018; ISBN 978-1-63081-348-2.
9. Palmeri, R.; Bevacqua, M.T.; Crocco, L.; Isernia, T.; Di Donato, L. Microwave imaging via distorted iterated virtual experiments. *IEEE Trans. Antennas Propag.* **2017**, *65*, 829–838. [CrossRef]
10. Solimene, R.; Buonanno, A.; Pierri, R. Imaging small PEC spheres by a linear δ approach. *IEEE Trans. Geosci. Remote Sens.* **2008**, *46*, 3010–3018. [CrossRef]
11. Gennarelli, G.; Vivone, G.; Braca, P.; Soldovieri, F.; Amin, M.G. Multiple extended target tracking for through-wall radars. *IEEE Trans. Geosci. Remote Sens.* **2015**, *53*, 6482–6494. [CrossRef]
12. Brancaccio, A.; Leone, G. Multimonostatic shape reconstruction of two-dimensional dielectric cylinders by a Kirchhoff-based approach. *IEEE Trans. Geosci. Remote Sens.* **2010**, *48*, 3152–3161. [CrossRef]
13. Anselmi, N.; Salucci, M.; Oliveri, G.; Massa, A. Wavelet-Based Compressive Imaging of Sparse Targets. *IEEE Trans. Antennas Propag.* **2015**, *63*, 4889–4900. [CrossRef]
14. Oliveri, G.; Randazzo, A.; Pastorino, M.; Massa, A. Electromagnetic imaging within the contrast-source formulation by means of the multiscaling inexact Newton method. *J. Opt. Soc. Am. A* **2012**, *29*, 945–958. [CrossRef] [PubMed]
15. D'Urso, M.; Isernia, T.; Morabito, A.F. On the solution of 2-D inverse scattering problems via source-type integral equations. *IEEE Trans. Geosci. Remote Sens.* **2010**, *48*, 1186–1198. [CrossRef]
16. Shumakov, D.S.; Nikolova, N.K. Fast quantitative microwave imaging with scattered-power maps. *IEEE Trans. Microw. Theory Tech.* **2018**, *66*, 439–449. [CrossRef]
17. Desmal, A.; Bağcı, H. Shrinkage-thresholding enhanced Born iterative method for solving 2D inverse electromagnetic scattering problem. *IEEE Trans. Antennas Propag.* **2014**, *62*, 3878–3884. [CrossRef]

18. Maaref, N.; Millot, P.; Ferrières, X.; Pichot, C.; Picon, O. Electromagnetic Imaging Method Based on Time Reversal Processing Applied to Through-the-Wall Target Localization. *Prog. Electromagn. Res.* **2008**, *1*, 59–67. [CrossRef]

19. Bellizzi, G.; Bucci, O.M.; Catapano, I. Microwave cancer imaging exploiting magnetic nanoparticles as contrast agent. *IEEE Trans. Biomed. Eng.* **2011**, *58*, 2528–2536. [CrossRef] [PubMed]

20. Catapano, I.; Di Donato, L.; Crocco, L.; Bucci, O.M.; Morabito, A.F.; Isernia, T.; Massa, R. On quantitative microwave tomography of female breast. *Prog. Electromagn. Res.* **2009**, *97*, 75–93. [CrossRef]

21. Miao, Z.; Kosmas, P. Multiple-frequency DBIM-TwIST algorithm for microwave breast imaging. *IEEE Trans. Antennas Propag.* **2017**, *65*, 2507–2516. [CrossRef]

22. Scapaticci, R.; Kosmas, P.; Crocco, L. Wavelet-based regularization for robust microwave imaging in medical applications. *IEEE Trans. Biomed. Eng.* **2015**, *62*, 1195–1202. [CrossRef] [PubMed]

23. Hopfer, M.; Planas, R.; Hamidipour, A.; Henriksson, T.; Semenov, S. Electromagnetic tomography for detection, differentiation, and monitoring of brain stroke: A virtual data and human head phantom study. *IEEE Antennas Propag. Mag.* **2017**, *59*, 86–97. [CrossRef]

24. Tournier, P.-H.; Bonazzoli, M.; Dolean, V.; Rapetti, F.; Hecht, F.; Nataf, F.; Aliferis, I.; El Kanfoud, I.; Migliaccio, C.; de Buhan, M.; et al. Numerical modeling and high-speed parallel computing: New perspectives on tomographic microwave imaging for brain stroke detection and monitoring. *IEEE Antennas Propag. Mag.* **2017**, *59*, 98–110. [CrossRef]

25. Bisio, I.; Estatico, C.; Fedeli, A.; Lavagetto, F.; Pastorino, M.; Randazzo, A.; Sciarrone, A. Brain stroke microwave imaging by means of a Newton-conjugate-gradient method in Lp Banach spaces. *IEEE Trans. Microw. Theory Tech.* **2018**, *66*, 3668–3682. [CrossRef]

26. Ireland, D.; Bialkowski, K.; Abbosh, A. Microwave imaging for brain stroke detection using Born iterative method. *Antennas Propag. IET Microw.* **2013**, *7*, 909–915. [CrossRef]

27. Bellizzi, G.G.; Crocco, L.; Cavagnaro, M.; Farina, L.; Lopresto, V.; Scapaticci, R. A full-wave numerical assessment of microwave tomography for monitoring cancer ablation. In Proceedings of the 11th European Conference on Antennas and Propagation (EUCAP), Paris, France, 19–24 March 2017; pp. 3722–3725.

28. Chandra, R.; Johansson, A.J.; Gustafsson, M.; Tufvesson, F. A microwave imaging-based technique to localize an in-body RF source for biomedical applications. *IEEE Trans. Biomed. Eng.* **2015**, *62*, 1231–1241. [CrossRef] [PubMed]

29. Boero, F.; Fedeli, A.; Lanini, M.; Maffongelli, M.; Monleone, R.; Pastorino, M.; Randazzo, A.; Salvadè, A.; Sansalone, A. Microwave tomography for the inspection of wood materials: Imaging system and experimental results. *IEEE Trans. Microw. Theory Tech.* **2018**, *66*, 3497–3510. [CrossRef]

30. Li, X.; Hagness, S.C. A confocal microwave imaging algorithm for breast cancer detection. *IEEE Microw. Compon. Lett.* **2001**, *11*, 130–132. [CrossRef]

31. Bozza, G.; Estatico, C.; Pastorino, M.; Randazzo, A. An inexact Newton method for microwave reconstruction of strong scatterers. *IEEE Antennas Wirel. Propag. Lett.* **2006**, *5*, 61–64. [CrossRef]

32. Lanini, M.; Poretti, S.; Salvade, A.; Monleone, R. Design of a slim wideband-antenna to overcome the strong reflection of the air-to-sample interface in microwave imaging. In Proceedings of the 2015 International Conference on Electromagnetics in Advanced Applications, Turin, Italy, 7–11 September 2015; pp. 1020–1023.

33. Whittow, W.G. 3D printing, inkjet printing and embroidery techniques for wearable antennas. In Proceedings of the 10th European Conference on Antennas and Propagation (EuCAP2016), Davos, Switzerland, 10–15 April 2016; pp. 1–4.

34. Bertero, M.; Boccacci, P. *Introduction to Inverse Problems in Imaging*; Institute of Physics Pub: Bristol, UK; Philadelphia, PA, USA, 1998; ISBN 978-0-7503-0439-9.

Journal of
Imaging

Article

Contraction Integral Equation for Three-Dimensional Electromagnetic Inverse Scattering Problems

Yu Zhong [1,*] **and Kuiwen Xu** [2]

[1]   Institute of High Performance Computing, Agency for Science, Technology and Research (A*STAR),
      Singapore 138632, Singapore
[2]   Key Lab of RF Circuits and Systems of Ministry of Education, Hangzhou Dianzi University,
      Hangzhou 310018, China; kuiwenxu@hdu.edu.cn
*   Correspondence: zhongyu@ihpc.a-star.edu.sg

Received: 31 December 2018; Accepted: 31 January 2019; Published: 8 February 2019

Abstract: Inverse scattering problems (ISPs) stand at the center of many important imaging applications, such as geophysical explorations, industrial non-destructive testing, bio-medical imaging, etc. Recently, a new type of contraction integral equation for inversion (CIE-I) has been proposed to tackle the two-dimensional electromagnetic ISPs, in which the usually employed Lippmann–Schwinger integral equation (LSIE) is transformed into a new form with a modified medium contrast via a contraction mapping. With the CIE-I, the multiple scattering effects, i.e., the physical reason for the nonlinearity in the ISPs, is substantially suppressed in estimating the modified contrast, without compromising physical modeling. In this paper, we firstly propose to implement this new CIE-I for the three-dimensional ISPs. With the help of the FFT type twofold subspace-based optimization method (TSOM), when handling the highly nonlinear problems with strong scatterers, those with higher contrast and/or larger dimensions (in terms of wavelengths), the performance of the inversions with CIE-I is much better than the ones with the LSIE, wherein inversions usually converge to local minima that may be far away from the solution. In addition, when handling the moderate scatterers (those the LSIE modeling can still handle), the convergence speed of the proposed method with CIE-I is much faster than the one with the LSIE. Secondly, we propose to relax the contraction mapping condition, i.e., different contraction mappings are used in updating contrast sources and contrast, and we find that the convergence can be further accelerated. Several numerical tests illustrate the aforementioned interests.

Keywords: inverse scattering; nonlinear problem; contraction integral equation for inversion (CIE-I); imaging

1. Introduction

Inverse scattering problems (ISPs) in electromagnetics and acoustics are of great interest in industries due to important imaging applications in various areas, such as geophysical survey, non-destructive testing, ground-penetrating radar, bio-medical imaging, etc., as solving the ISPs provides rich information about the unknown targets, such as locations, shapes, and material distributions within some structures [1–4]. For instance, microwave imaging has been used to inspect the abnormalities in human bodies like bleeding in the head and tumours in breasts. On the other hand, they are also quite important due to the representative difficulties in solving a large group of inverse problems concerning waves and fields, and thus researchers in mathematics, physics, and engineering societies have devoted great efforts to improving efficiencies and accuracies of the numerical solvers [5]. As shown in Figure 1, solving ISPs is to determine the unknown scatterers within a domain, when the domain is illuminated by several different incidences and we can measure the scattered fields outside the domain (usually contaminated by noise) for each incidence. It is well known that the main difficulties in such ISPs are their two intrinsic properties, i.e., ill-posedness and nonlinearity [1]. As most of the practical problems are three-dimensional (3-D) ones, which require much more computational resources than those in two-dimensional cases, they are often more difficult to solve due to the data deficiency (measurement aperture usually covers a small solid angle), and therefore also stronger ill-posedness and nonlinearity. Great efforts have been paid to tackle these demanding problems, for instance [6–9].

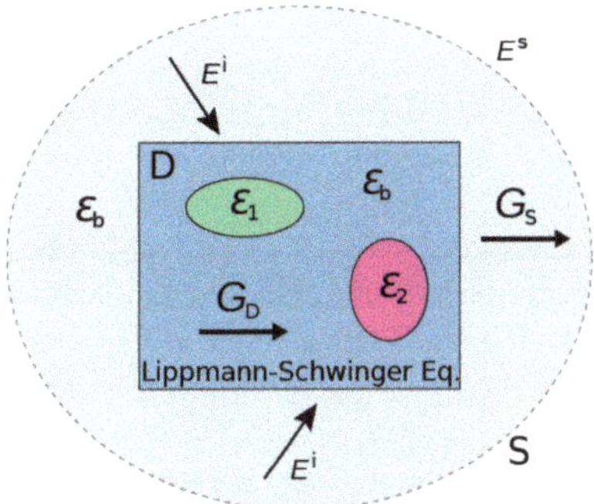

Figure 1. Schematics of inverse scattering problems.

Methods for solving ISPs can be catalogued into two types of optimization approaches: deterministic and stochastic. The deterministic type of the inversion methods has been developed for decades, such as the contrast source inversion (CSI) method [6,10,11], the Born iterative method and distorted Born iterative method [12,13], the level set method [14,15], the subspace-based optimization method (SOM) [8,16–18], etc. The second type, the stochastic type of the inversion methods, usually employs a group of initial guesses and uses the stochastic optimization scheme to minimize the objective function, such as the genetic algorithm and the evolutionary optimization. The stochastic type methods increase the possibilities of finding the global minimum rather than being trapped in a local minimum as the deterministic optimization techniques [19,20]. Both techniques have also been applied to solve 3-D inverse scattering problems [6,21–24]. Other than the quantitative methods mentioned above, some qualitative methods, such as the linear sampling method [25] and other methods, like [26], are proposed to retrieve the geometric supports of the unknown scatterers.

To tackle the nonlinearity, the major difficulty in efficiently solving the ISPs, different types of integral equations have been proposed, such as in [27–29]. In particular, motivated by the contraction integral equation (CIE) in solving the direct scattering problems with highly conductive background media [30], a new type of contraction integral equation for inversion (CIE-I) has been proposed [29] to tackle the two-dimensional highly nonlinear ISPs by transforming the usually employed Lippmann–Schwinger integral equation (LSIE) into a new form with a modified medium contrast via a contraction mapping. With the CIE-I, the global multiple scattering effects (MSE) are substantially suppressed in estimating the modified contrast in the CSI type methods. With the FFT type twofold SOM regularization scheme [8], the inversion solver with CIE-I is capable of effectively alleviating the nonlinearity of the two-dimensional ISPs, especially those highly nonlinear ones with strong scatterers with large contrast and/or large dimensions (in terms of wavelengths). In this paper, this new CIE-I will be implemented in tackling the 3-D ISPs.

In summary, the contributions of the paper include:

1. The CIE-I is firstly implemented to tackle the computationally costly full 3-D ISPs, to address the highly nonlinear 3-D ISPs, and to accelerate the convergence of the inversions.
2. A relaxed type of inversion scheme based on CIE-I is proposed, with different auxiliary parameters β (the parameter in CIE-I to control the portion of MSE in estimating the contrast) in

updating the contrast sources and in updating the contrast. This means to further accelerate the convergence of the inversions.

3. Several numerical tests are provided with details, for the sake of further algorithmic studies.

After the Introduction, the proposed 3-D inversion method is detailed in Section 2. In Section 3, three numerical examples are given to validate the proposed method. Finally, conclusions are drawn.

2. Inversion with CIE-I

In this paper, the domains of interest (DoI) are chosen to be rectangular cuboid in order to implement the conjugate gradient fast Fourier transform (CG-FFT) scheme and apply the Fourier basis in TSOM. For the convenience of reading, in this paper, we denote the one-dimensional tensor as $\bar{a}$, two-dimensional tensor as $\bar{\bar{a}}$, three-dimensional tensor as $\hat{a}$, and four-dimensional tensor as $\hat{a}$. Unless otherwise specified, the subscript of the tensors denotes the index of the element, such as $\bar{\bar{a}}_{m,n}$ denotes the element in $\bar{\bar{a}}$ with index $\{m,n\}$. We use bold symbols to denote vectorial physical quantities, such as the positions r and the electric fields E in 3-D cases.

2.1. 3-D Modeling

In 3-D cases, there are N_i incident waves from different angles impinging onto the rectangular cuboid DoI $\mathcal{D}$ ($\mathcal{D} \subset \mathcal{R}^3$, the background 3-D homogeneous medium with permittivity ϵ_0 and permeability μ_0), where nonmagnetic scatterers are located, and these incident waves are expressed as $E_l^{\mathrm{inc}}(r), l = 1, 2, \ldots, N_i, r \in \mathcal{D}$. For each incidence, the scattered fields are collected by N_r receivers located at $r'_j, j = 1, 2, \ldots, N_r$. With all such information, including every incident field inside the domain of interest and the corresponding scattered fields at the positions of all detectors, we aim at determining the dielectric profile $\epsilon(r), r \in \mathcal{D}$.

The scattering models are governed by the following electric field volume integral equation based on the well-known Lippmann–Schwinger integral equation (LSIE). For the lth incidence, the field equation in the domain $\mathcal{D}$ is expressed as

$$I_l(r) = \chi(r)E_l^{\mathrm{inc}}(r) + \chi(r)(G_{\mathrm{D}}^{3\mathrm{D}}I_l)(r), \tag{1}$$

where $\chi(r) = (\epsilon_{\mathrm{r}}(r) - 1)$ is the contrast, $\epsilon_{\mathrm{r}}(r)$, $I_l(r) = \chi(r)E_l^{\mathrm{tot}}(r)$ and $E_l^{\mathrm{inc}}(r)$ are the relative permittivity, the contrast source and the incident electric field at r, respectively, and $(G_{\mathrm{D}}^{3\mathrm{D}}I_l)(r)$ is an integral operator with the dyadic Green's function as the integral kernel, which can be written as [8,31],

$$(G_{\mathrm{D}}^{3\mathrm{D}}I_l)(r) := k_0^2 \int_{\mathcal{D}} \left(\bar{\bar{I}} + \frac{\nabla\nabla}{k_0^2}\right) \cdot g(r,r')I_l(r')dr', \tag{2}$$

with $k_0 = \omega\sqrt{\epsilon_0\mu_0}$ as the wave number of the background medium, and $g(r,r')$ as the 3-D Green's function for the background homogeneous medium. Notice that the $I_l(r)$ is the contrast source in this paper, while, in [8], it is the physically induced current that includes a multiplicative factor $-i\omega\epsilon_0$ compared to the contrast source. Otherwise, as we know, the dyadic Green's function is composed of nine scalar elements, which represent the mapping from the three components of the contrast source to the three components of the scattered fields inside the DoI.

The nonlinearity of ISPs comes from the MSE, as shown in Equation (1). To alleviate the nonlinearity of the model, in [29], by some mathematical manipulation on Equation (1), another new-type integral equation, which is denoted as CIE-I herein, can be obtained

$$\beta(r)I_l(r) = R(r)\beta(r)I_l(r) + R(r)\left[E_l^{\mathrm{inc}}(r) + (G_{\mathrm{D}}^{3\mathrm{D}}I_l)(r)\right], \tag{3}$$

where $R(r) = \beta(r)\chi(r)/[\beta(r)\chi(r) + 1]$ is the modified contrast function, and $\beta(r)$ is a chosen auxiliary parameter to control the portion of the MSE in estimating the contrast [29], which can be a constant or a variable at the different position in the DoI. For the convenience of discretizing the equation, we rewrite the dyadic Green's function as

$$(G^{3D}_{D;uv}I_{l;v})(r) := \begin{cases} k_0^2(1 + \dfrac{1}{k_0^2}\dfrac{\partial^2}{\partial \iota_u^2}) \int_D g(r,r')I_{l;v}(r')dr', & \text{if } u = v, \\[2ex] \dfrac{\partial^2}{\partial \iota_u \partial \iota_v} \int_D g(r,r')I_{l;v}(r')dr', & \text{if } u \neq v, \end{cases} \tag{4}$$

where $u, v = 1, 2, 3$ represent the x-, y- and z-components of a vector, respectively, and $\iota_1 = x$, $\iota_2 = y$, and $\iota_3 = z$.

Following the conventions in [8], we can rewrite the vectorial Equation (3) into three coupled scalar equations in the discrete forms. Thus, we first discretize the rectangular domain of interest into small cuboid subdomains, whose dimensions are much smaller than the wavelength and the center of which are located at $r_{m,n,p}$, with m, n, p integers and $m \in [1, M_1]$, $n \in [1, M_2]$ and $p \in [1, M_3]$. Here, M_1, M_2, and M_3 are the total number of subdomains along x-, y-, and z-directions, respectively, and we let $M = M_1 \times M_2 \times M_3$ be the total number of the subdomains. With such discretization, we have

$$\hat{\beta}_{m,n,p}\hat{I}_{l;u;m,n,p} = \hat{R}_{m,n,p}\hat{\beta}_{m,n,p}\hat{I}_{l;u;m,n,p} + \hat{R}_{m,n,p}\left[\hat{E}^{inc}_{l;u;m,n,p} + \sum_{v=1}^{3} \hat{G}^{3D}_{D;uv;m,n,p}(\hat{I}_{l;v})\right], \tag{5}$$

where $\hat{R}_{m,n,p} = \hat{\beta}_{m,n,p}\hat{\chi}_{m,n,p}/[\hat{\beta}_{m,n,p}\hat{\chi}_{m,n,p} + 1]$, $\hat{\chi}_{m,n,p}$ is the contrast at $r_{m,n,p}$, whereas $\hat{E}^{inc}_{l;u;m,n,p}$ and $\hat{I}_{l;u;m,n,p}$ are the incident electric field and the induced current at $r_{m,n,p}$, respectively. Subscript $u = 1, 2, 3$ denotes the x, y, and z components of a vector. Note that the convolution-type operators $\hat{G}^{3D}_{D;uv}$ are obtained via the Equation (4). For further details of the discretization of Equation (1) and the finite difference scheme to generate $(G^{3D}_D I_l)(r)$, please refer to the Appendix of [6].

Similarly, the integral operator relating the contrast sources and the scattered fields could also be expressed as the summation of the contribution from all the subdomains,

$$\overline{E}^{sca}_l = \overline{\overline{G}}^{3D}_S \cdot \overline{I}_l, \tag{6}$$

where $\overline{E}^{sca}_l = \left[\overline{E}^{sca T}_{l;1}, \overline{E}^{sca T}_{l;2}, \overline{E}^{sca T}_{l;3}\right]^T$ is a $3N_r$ dimensional vector with $\overline{E}^{sca}_{l;\kappa} = \left[E^{sca}_{l;\kappa;1}, E^{sca}_{l;\kappa;2}, \ldots, E^{sca}_{l;\kappa;N_r}\right]^T$ ($\kappa = 1, 2, 3$ denotes the x, y, z component of the corresponding vector, respectively), $\overline{I}_l$ is a $3M$ dimensional vector obtained by $\overline{I}_l = \text{vec}\left\{\hat{I}_l\right\}$. In a 3-D scenario, the vectorization operation $\text{vec}\{\cdot\}$ is defined to vectorize a four-dimensional tensor into a vector, i.e., if $\overline{I}_l = \text{vec}\left\{\hat{I}_l\right\}$, we have $\overline{I}_{l;\vartheta} = \hat{I}_{l;m,n,p,\kappa}$ with $\vartheta = (\kappa - 1) \times M + (p - 1) \times (M_1 \times M_2) + (n - 1) \times M_1 + m$. In Equation (6), the scattering operator is defined as

$$\overline{\overline{G}}^{3D}_S = \begin{bmatrix} \overline{\overline{G}}_{S;11} & \overline{\overline{G}}_{S;12} & \overline{\overline{G}}_{S;13} \\ \overline{\overline{G}}_{S;21} & \overline{\overline{G}}_{S;22} & \overline{\overline{G}}_{S;23} \\ \overline{\overline{G}}_{S;31} & \overline{\overline{G}}_{S;32} & \overline{\overline{G}}_{S;33} \end{bmatrix}, \tag{7}$$

a $3N_r \times 3M$ matrix, with $\overline{\overline{G}}_{S;uv}$, a $N_r \times M$ matrix, the mapping from the v component of the induced current to the u component of scattered fields (the subscripts $u, v = 1, 2, 3$ are not indexed for tensor elements). The explicit expression of $\overline{\overline{G}}_{S;uv}$ is

$$\overline{\overline{G}}^{3D}_{S;uv}(a,b) = \left\{ \left(k_0^2 + \frac{ik_0}{R_{a,b}} - \frac{1}{R_{a,b}^2}\right)\delta(u - v) + \right. \\ \left. [(r'_a)_u - (r_{m,n,p})_u][(r'_a)_v - (r_{m,n,p})_v]\left(-\frac{k_0^2}{R_{a,b}^2} - \frac{3ik_0}{R_{a,b}^3} + \frac{3}{R_{a,b}^4}\right)\right\} g(r'_a, r_{m,n,p}), \tag{8}$$

where $\delta(y)$ is 1 when $y = 0$ and is 0 otherwise, $R_{a,b} = \left| r'_a - r_{m,n,p} \right|$ in which $b = (p-1) \times (M_1 \times M_2) + (n-1) \times M_1 + m$ with $a = 1, 2, \ldots, N_r$, $b = 1, 2, \ldots, M$, $m = 1, 2, \ldots, M_1$, $n = 1, 2, \ldots, M_2$, and $p = 1, 2, \ldots, M_3$. Here, $(r)_u$ denotes the u component of r.

2.2. Objective Function for Inversions

In this subsection, we build the objective function used in the proposed inversion method, in which we will use the FFT-TSOM [8] as the regularization to stabilize the inversion, as done in [29]. Emphasize that the twofold subspace constraints have different regularization effects. The first fold, the original SOM, balances the two mismatches in the objective function, whereas the second fold, the TSOM, is the key to stabilize the inversions with CIE-I. Details are given in below.

For the original SOM part, as given in [16], with the spectral information of $\overline{\overline{G}}_S^{3D}$ (the singular value decomposition–SVD of $\overline{\overline{G}}_S^{3D}$ tells $\overline{\overline{G}}_S^{3D} \cdot \overline{v}_j^S = \sigma_j^S \overline{u}_j^S$ and the complexity of thin SVD of $\overline{\overline{G}}_S^{3D}$ is $O(27N_r^2 M)$, assuming that the singular values σ_j^S is a non-increasing sequence), the contrast sources can be decomposed into two parts, deterministic part of the contrast sources (DPCS) and ambiguous part of the contrast sources (APCS), the former being obtained as

$$\overline{I}_l^d = \sum_{j=1}^{L} \frac{\overline{u}_j^{S*} \cdot \overline{E}_l^{sca}}{\sigma_j^S} \overline{v}_j^S = \overline{\overline{V}}_S^+ \cdot \overline{\alpha}_l^+, \tag{9}$$

where $\overline{\overline{V}}_S^+ = \left[\overline{v}_1^S, \overline{v}_2^S, \ldots, \overline{v}_L^S \right]$, $\overline{\alpha}_l^+ = \left[\alpha_{l;1}^+, \alpha_{l;2}^+, \ldots, \alpha_{l;L}^+ \right]^T$ with $\alpha_{l;j}^+ = (\overline{u}_j^{S*} \cdot \overline{E}_l^{sca})/\sigma_j^S$, $j = 1, 2, \ldots, L$, and the superscript $*$ denotes the Hermitian operation while superscript $+$ refers to the dominant current subspace, the subspace corresponding to the dominant singular values. The value of L is chosen according to the noise level [16]. Later, we will see how this might work after introducing the objective function for the inversion.

For the second fold subspace constraint, according to the FFT-TSOM [8], the APCS can be written as

$$\overline{I}_l^a(\hat{\overline{\gamma}}_l) = \overline{I}_l^{tmp} - \overline{\overline{V}}_S^+ \cdot \left(\overline{\overline{V}}_S^{+*} \cdot \overline{I}_l^{tmp} \right), \tag{10}$$

where $\hat{\overline{\gamma}}_l = \left[\hat{\gamma}_{x;l}; \hat{\gamma}_{y;l}; \hat{\gamma}_{z;l} \right]$, $\overline{I}_l^{tmp} = \left[\overline{I}_{x;l}^{tmp}; \overline{I}_{y;l}^{tmp}; \overline{I}_{z;l}^{tmp} \right]$ with

$$\overline{I}_{u;l}^{tmp} = \text{vec}\left\{ \text{IDFT}\left\{ \hat{\gamma}_{u;l} \right\} \right\} \tag{11}$$

and $u = x$, y, and z, $\text{IDFT}\{\}$ is the inverse discrete Fourier transform operator, the $\text{vec}\{\cdot\}$ is the vectorization operator. Note that the inverse discrete Fourier transform (IDFT) is performed by the 3-D FFT algorithm, the computational complexity of which is $O(M \log_2 M)$, with $M = M_1 \times M_2 \times M_3$.

By using the low-frequency Fourier components, we are able to constrain the APCS within a low-dimensional subspace. The reason [17] is to use only the contrast sources components in this subspace that is influential to the scattered fields within the DoI, such that the stability of the inversions is substantially increased. To reduce the computational costs, such a subspace can be approximately spanned by low-frequency Fourier bases [8]. Here, if we use all Fourier bases, the construction of the APCS becomes the one in the original SOM. Otherwise, we can use the low-frequency components, and set the coefficients for those high-frequency components as zeros. This can be achieved via using a mask with eight corners being 1, with size M_F ($1 \leq M_F \leq M_1/2, M_2/2, M_3/2$), and other positions being 0. The details can be found in [8].

Having expressed the contrast sources as aforementioned, it is straightforward to define the objective function. Firstly, it is natural to give the mismatch of the scattered fields by

$$\Delta_l^{fie}(\hat{\overline{\gamma}}_l) = \left\| \overline{\overline{G}}_S \cdot \overline{I}_l^a + \overline{\overline{G}}_S \cdot \overline{I}_l^d - \overline{E}_l^{sca} \right\|^2, \tag{12}$$

where $\overline{I}_l^{\mathrm{d}}$ and $\overline{I}_l^{\mathrm{a}}$ are as in Equations (9) and (10), respectively, and $\|\cdot\|$ denotes the L^2 norm of a tensor. The current equation in Equation (5) is another key equation to satisfy. Using the APCS construction Equation (10), we define an operator as

$$\left(\mathcal{L}^{3D}\left(\hat{\gamma}_l\right)\right)_{\kappa;m,n,p} = \beta \hat{I}^{\mathrm{a}}_{l;\kappa;m,n,p} - \beta \hat{R}_{m,n,p} \hat{I}^{\mathrm{a}}_{l;\kappa;m,n,p} - \hat{R}_{m,n,p}\left[\sum_{v=1}^{3} \hat{G}^{3D}_{D;\kappa v;m,n,p}(\hat{I}^{\mathrm{a}}_{l;v})\right]. \tag{13}$$

With this definition, we could write the mismatch of Equation (5) as

$$\Delta_l^{\mathrm{cur}}(\hat{\gamma}_l, \hat{R}) = \left\|\mathcal{L}^{3D}(\hat{\gamma}_l) - \hat{I}_l^{3D}\right\|^2, \tag{14}$$

where $\hat{I}^{3D}_{l;\kappa;m,n,p} = -\beta \hat{I}^{\mathrm{d}}_{l;\kappa;m,n,p} + \beta \hat{R}_{m,n,p} \hat{I}^{\mathrm{d}}_{l;\kappa;m,n,p} + \hat{R}_{m,n,p}\left[\left\{\sum_{v=1}^{3} \hat{G}^{3D}_{D;\kappa v;m,n,p}(\hat{I}^{\mathrm{d}}_{l;v})\right\} + \hat{E}^{\mathrm{inc}}_{l;\kappa;m,n,p}\right].$ Then, the objective function can be given as

$$f(\hat{\gamma}_1, \hat{\gamma}_2, \dots, \hat{\gamma}_{N_i}, \hat{R}) = \sum_{l=1}^{N_i} \left(\Delta_l^{\mathrm{fie}} / \left\|\overline{E}_l^{\mathrm{sca}}\right\|^2 + \Delta_l^{\mathrm{cur}} / \left\|\overline{E}_l^{\mathrm{inc}}\right\|^2\right). \tag{15}$$

The inversion is to minimize this objective function. As in [8], the conjugate gradient (CG) type algorithm that is used in contrast source inversion (CSI) method is adopted to minimize this nonlinear problem by alternatively updating the $\hat{\gamma}_l$ and $\hat{R}$ at every iteration of the optimization.

2.3. Sketch of the Inversion Method

Following the inversion method in [8,29], we summarize it as follows:

1. Set the background medium and null APCS as the initial guesses and choose an L value such that the corresponding first L singular values of $\overline{\overline{G}}_S^{3D}$ are larger than the noise level (assuming that the noise is a white Gaussian one).
2. Set proper values for β in CIE-I modeling and proper value for M_F to control the number of Fourier bases being used.
3. Carry out the CG type optimization algorithm to alternatively update the two types of variables, where the APCS is updated with a one-step Polak–Ribière CG scheme and the contrast is updated with the least squares method.
4. Stop the optimization if a termination condition is met, which can be a maximum number of iterations or a pre-defined relative change of APCS coefficients.
5. If the maximum number of rounds of inversion is met, go to Step 6. Otherwise, the obtained contrast and APCS will be used as the initial guesses for the next round of optimization with smaller β and larger M_F in Step 3.
6. Output the obtained contrast.

As mentioned in [29], the first round inversion with a large β and with a low-dimensional subspace enables a fast convergence to a meaningful coarse result that could be used as an initial guess in the second round. By increasing the dimension of the APCS subspace and including more MSE in estimating the contrast, the inversion gives us a result with better resolution. Since in the second round a (supposedly) good initial guess is given, the convergence is also very fast. For some difficult problems, the inversion could be carried out with multi-round optimizations by gradually increasing the dimensions of the APCS subspace in each round while decreasing the values of β in estimating the contrast.

2.4. Updating Contrast and Contrast Sources with Different β

As mentioned above, there are two updates at each iteration for the two types of unknowns. In [29], the same β is used in each round of optimization for both updates. In the first round optimization in inversions, to tackle the nonlinearity, a large β is needed in updating the modified contrast, and therefore we use the same β for the update of APCS. In the subsequent rounds (instead of the first round) of optimization, for the sake of stability, a small β is used in estimating the modified contrast to cope with a high-dimensional APCS subspace, as discussed in [29]. However, there is not such a need to use a small β in updating the APCS. Consequently, in the second round of optimization, we can use a CIE-I model with larger value β_1 for the update of the APCS and another with smaller value β_2 for the update of the contrast at the same iteration. The same applies to the third round optimization, if there is such a need. The purpose of doing so is to further accelerate the convergence of these computationally burdensome 3-D inversions after the first round optimization.

As for the value range for β_2, we carry out a similar calculation as in [29], which reflects the norm of the scattering operator that maps the contrast sources to the scattered fields within DoI. The results shown in Figure 2 indicate that, if we want to suppress the MSE in estimating the contrast when using the least squares method, we might need to choose a value for β_2 that is larger than 3.5, and this is the guideline for the first round optimization. For the second or subsequent rounds, as good initial guesses are provided, we can include more MSE to retrieve the fine features of the unknown scatterers.

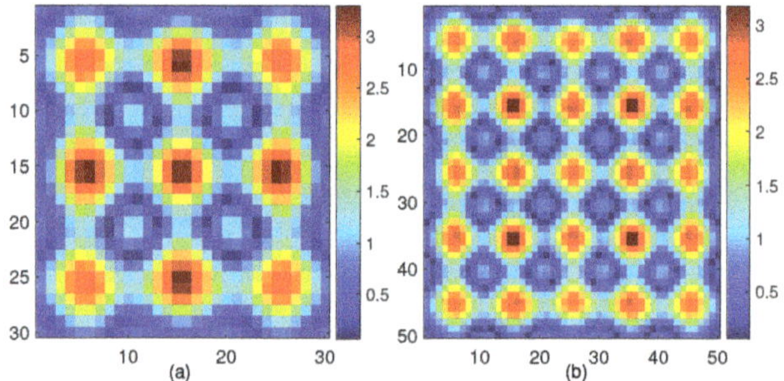

Figure 2. The values of $\left|(G^{3D}_{D;xx} I_x)(r)\right|$ in the plane $x = 0$ when $I_x = 1$ for a DoI with size (**a**) $3\lambda \times 3\lambda \times 3\lambda$ and (**b**) $5\lambda \times 5\lambda \times 5\lambda$.

3. Numerical Simulations

In this section, we will test the proposed inversion methods in three examples. We will use the same physical setup of sources and receivers for the 3-D example in [8], as shown in Figure 3. The DoI are all the same, a box with size $3\lambda \times 3\lambda \times 3\lambda$. The box is illuminated by 60 electric dipole antennas at 300 MHz ($\lambda = 1$ m in air), located at three circles (with 20 dipole antennas evenly distributed on each) with the same radius 3 m. The three circles are in $x - y$, $y - z$ and $x - z$ planes, and their centers are at (0.2, 0, –0.1), (0.1, 0, –0.15), and (–0.05, 0.1, 0), respectively. The direction of the electric dipole sources in the $y - z$ plane are in the x-direction, while those in the $x - z$ and $x - y$ planes are in the y- and z-directions, respectively. Scattered fields are measured by 60 receivers, located at the same positions as the 60 dipole sources. We collect all three components of the fields. Consequently, we have a 60×180 synthetic data matrix. In all three tests, the synthetic data are calculated with a $60 \times 60 \times 60$ mesh, while, in the inversions, a $30 \times 30 \times 30$ mesh is used. All synthetic data are contaminated with additive Gaussian white noise with level of 10%.

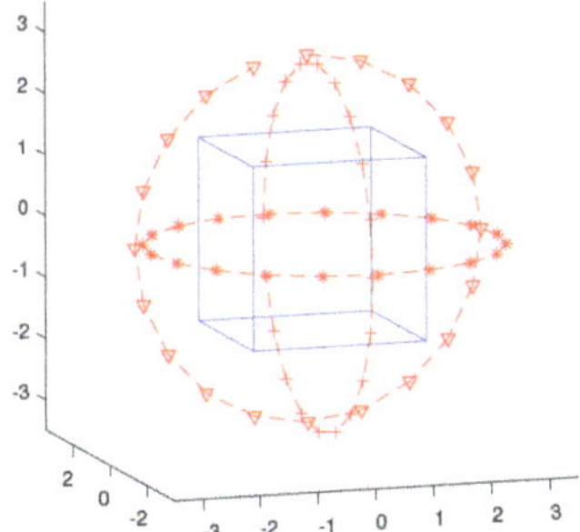

Figure 3. The physical setup for the numerical tests.

The termination condition in each round of optimization is to reach a pre-defined relative changing rate of the APCS coefficients, given as

$$\delta^{3D} = \sqrt{\frac{1}{N_i} \sum_{l=1}^{N_i} \frac{\|\hat{\gamma}_l^{(h)} - \hat{\gamma}_l^{(h-1)}\|^2}{\|\hat{\gamma}_l^{(h-1)}\|^2}}, \tag{16}$$

where $\hat{\gamma}_l^{(h)}$ is the APCS Fourier coefficient for the lth incidence at h^{th} iteration. We set $\delta^{3D} < 10^{-3}$ as the termination condition.

The reconstruction results are quantitatively evaluated by the following error:

$$Err = \sqrt{\frac{1}{M} \sum_{m,n,p} \frac{|\hat{\epsilon}_{r;m,n,p}^{est} - \hat{\epsilon}_{r;m,n,p}^{tr}|^2}{|\hat{\epsilon}_{r;m,n,p}^{tr}|^2}}. \tag{17}$$

When running on a workstation with eight threads and 32 Gb RAM, every iteration of the proposed inversion method costs about 25 seconds CPU time in MATLAB, including one update on the contrast sources and one update on the contrast. Due to the independence between different incidences, we use 8 MATLAB workers to update the contrast sources in parallel.

3.1. Example 1

We reuse the 3-D test in [8], where a coated cube is employed. The cube is with an outer layer ($\epsilon_{r2} = 1.5 + i0.3$) and an inner layer ($\epsilon_{r1} = 2 + i0.8$), where the outer edge length is $b = 2\lambda$ and inner edge length is $a = 1\lambda$, as shown in Figure 4a. In [8], we observe that the inversion with LSIE and FFT-TSOM is able to satisfactorily retrieve the cube while the LSIE with SOM fails. Here, we illustrate how the proposed inversion method with CIE-I and FFT-TSOM performs. Similarly, we carry out two rounds of optimization for this inversion as done in [8] with $M_F = 6$ and 10, but with CIE-I, we use $\beta = 6$ in the first round and $\beta = 1$ for the second round. The reconstruction results are shown in Figure 5, with $\Re\{\hat{\epsilon}_r^{est}\}$ as the real part of the reconstructed relative permittivity and $\Im\{\hat{\epsilon}_r^{est}\}$ as the imaginary part, where we see after two rounds of optimization that the coated cube is successfully found.

Now, we compare the convergence speeds of inversions with LSIE and CIE-I. We plot the reconstruction errors for both cases, as shown in Figure 6, in which the inversion with CIE-I appears to be converging much faster than the one with LSIE, with the same reconstruction quality. This confirms again the speeding convergence of inversions in two-dimensional cases shown in [29]. We would like to emphasize that this fast convergence property is extremely important in handling the 3-D ISPs, since the computational costs in 3-D problems are much higher than in two-dimensional cases.

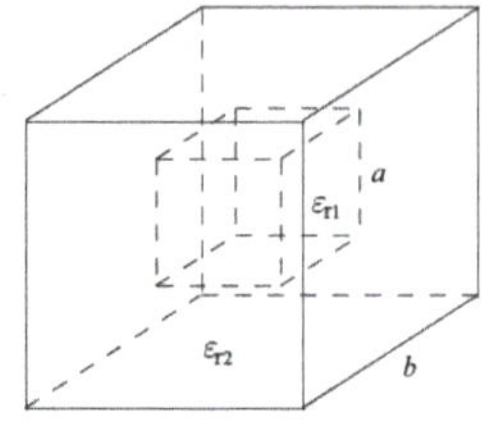

(**a**) The coated cube in *Example 1*.

(**b**) Real part of the true profile of the coated cube.

(**c**) Imaginary part of the true profile of the coated cube.

Figure 4. Coated cube used in *Example 1*.

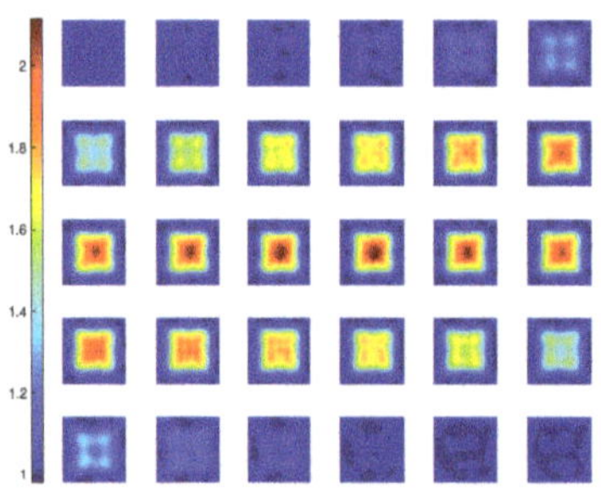

(**a**) $\Re\{\hat{\epsilon}_r^{\text{est}}\}$ after the first round optimization.

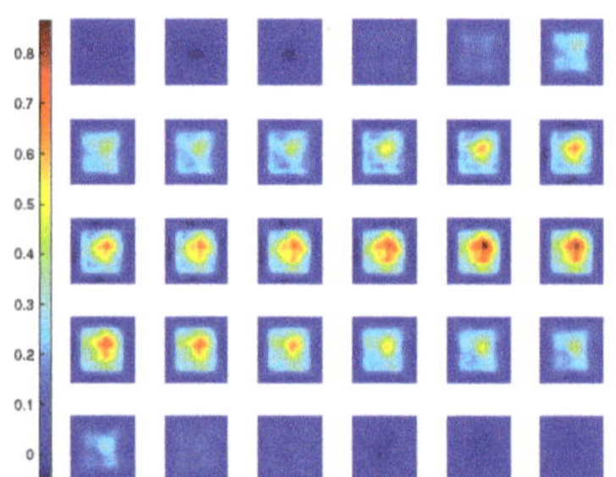

(**b**) $\Im\{\hat{\epsilon}_r^{\text{est}}\}$ after the first round optimization.

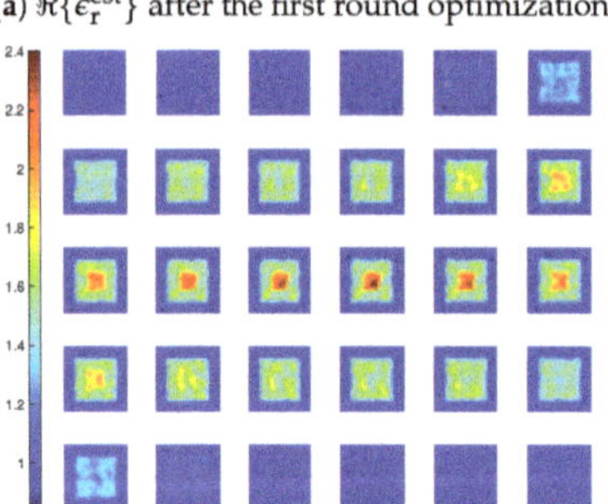

(**c**) $\Re\{\hat{\epsilon}_r^{\text{est}}\}$ after the second round optimization.

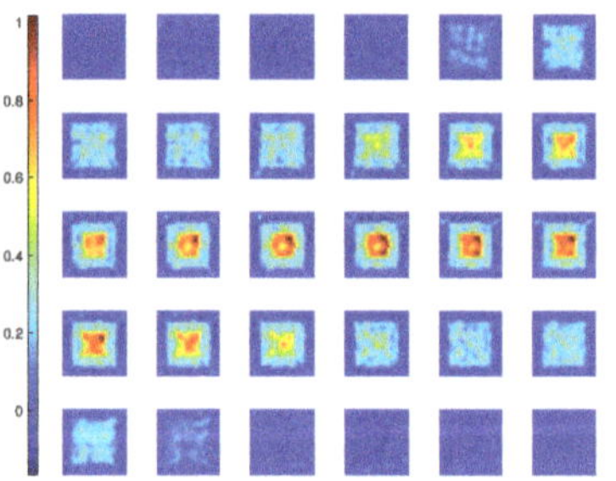

(**d**) $\Im\{\hat{\epsilon}_r^{\text{est}}\}$ after the second round optimization.

Figure 5. Reconstruction results of inversions with CIE-I. In the four subfigures, the 30 slices of DoI with each at $z = z_q$ plane, where z_q, $q = 1, \ldots, 30$, are the grid points along z-direction, and $z_p < z_q$ if $p < q$. The displaying sequence is with the convention of left to right, and top to down. For instance, the top left corner one is with z_1, and the top row second column one is with z_2. The same applies to the figures hereafter. In the first round, $\beta_1 = \beta_2 = 6$. In the second round, $\beta_1 = \beta_2 = 1$.

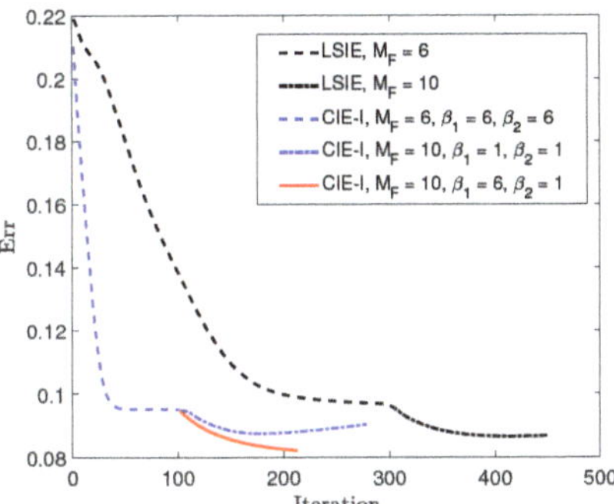

Figure 6. Errors of reconstructions obtained by different inversion methods for *Example 1*.

With this example, we further test the relaxed scheme with different β for updating the APCS and the contrast. The termination condition remains the same, $\delta^{3D} < 10^{-3}$. We apply this scheme in the second round optimization, with $\beta_1 = 6$ for the APCS update and $\beta_2 = 1$ for the contrast update. For comparison, the recorded reconstruction error at each iteration is plotted in Figure 6 as well, which shows that the convergence is further accelerated compared with the case of using $\beta_1 = 1$.

3.2. Example 2

In this test, we use a profile with four cubes with increasing contrasts. Having edge length 0.8λ, they are centered at $(-0.7, 0.7, -0.7)\lambda$, $(0.7, -0.7, -0.7)\lambda$, $(-0.7, -0.7, 0.7)\lambda$, and $(0.7, 0.7, 0.7)\lambda$, with $\epsilon_r = 2, 3, 4$ and 5, respectively, as shown in Figure 7.

Figure 7. True permittivity profile (real part) for *Example 2*.

Two inversions are carried out with LSIE and CIE-I, each with two rounds ($M_F = 6$ and 10). For CIE-I, the choices of β are 6 and 1 in the two rounds. The reconstruction results and errors are shown in Figures 8–10. From these results, we clearly observe that the inversion with LSIE is only able to find the bottom two cubes with lower contrasts, whereas the inversion with CIE-I succeeds in finding all four of four cubes. Reconstruction errors displayed in Figure 9 confirm it, inversion with LSIE diverging while converging with CIE-I.

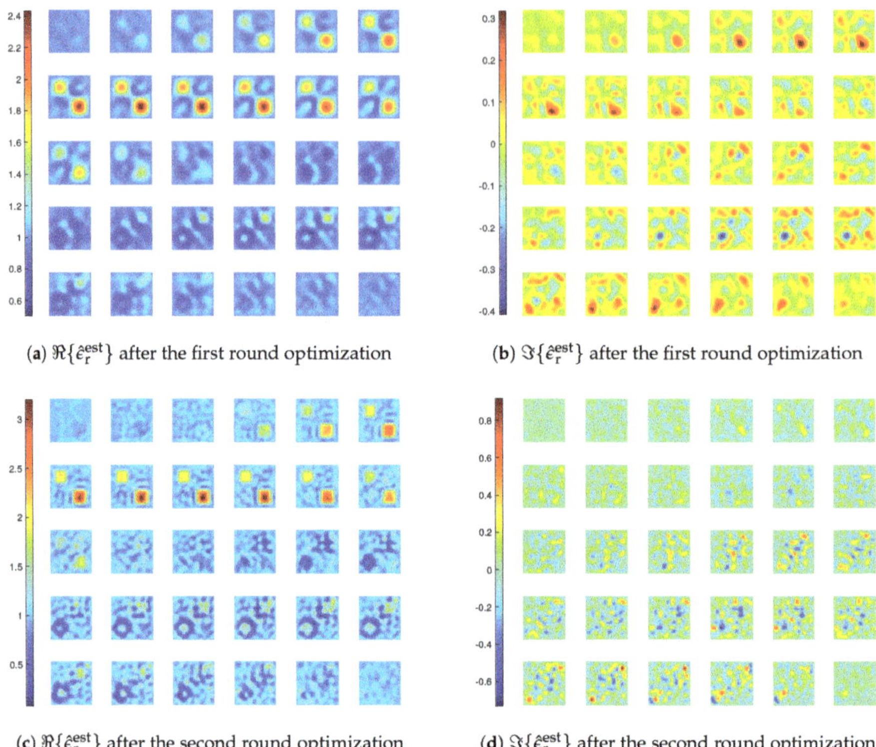

(**a**) $\Re\{\hat{\epsilon}_r^{est}\}$ after the first round optimization

(**b**) $\Im\{\hat{\epsilon}_r^{est}\}$ after the first round optimization

(**c**) $\Re\{\hat{\epsilon}_r^{est}\}$ after the second round optimization

(**d**) $\Im\{\hat{\epsilon}_r^{est}\}$ after the second round optimization

Figure 8. Reconstruction results of inversions with LSIE for *Example 2*.

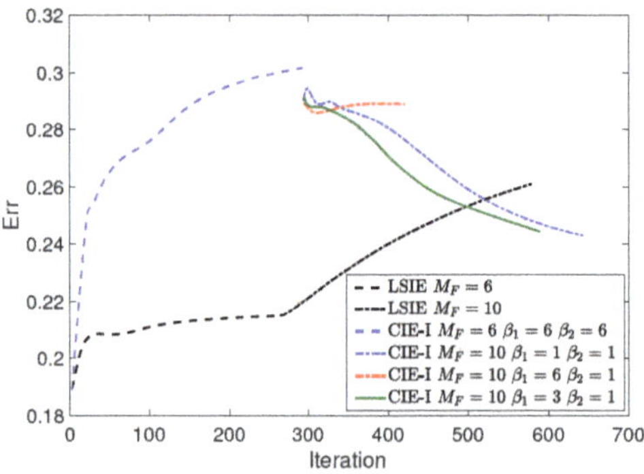

Figure 9. Errors of reconstructions obtained by different inversion methods for *Example 2*.

In addition, we test the inversion scheme with different β for updating the APCS and for updating the contrast. Firstly, we choose $\beta_1 = 6$ for the update of APCS and $\beta_2 = 1$ for the update of the contrast in the second round of optimization with $M_F = 10$, which however immaturely converges. This might be due to the reason of using a large β_1. Then, we choose $\beta_1 = 3$, and the optimization converges to a good solution with a convergence speed faster than when using $\beta_1 = 1$.

From this example, the inversion with CIE-I is able to handle the highly nonlinear problems consisting of strong scatterers, those with either large contrasts or large dimensions (in terms of wavelength), while the inversion with LSIE may fail.

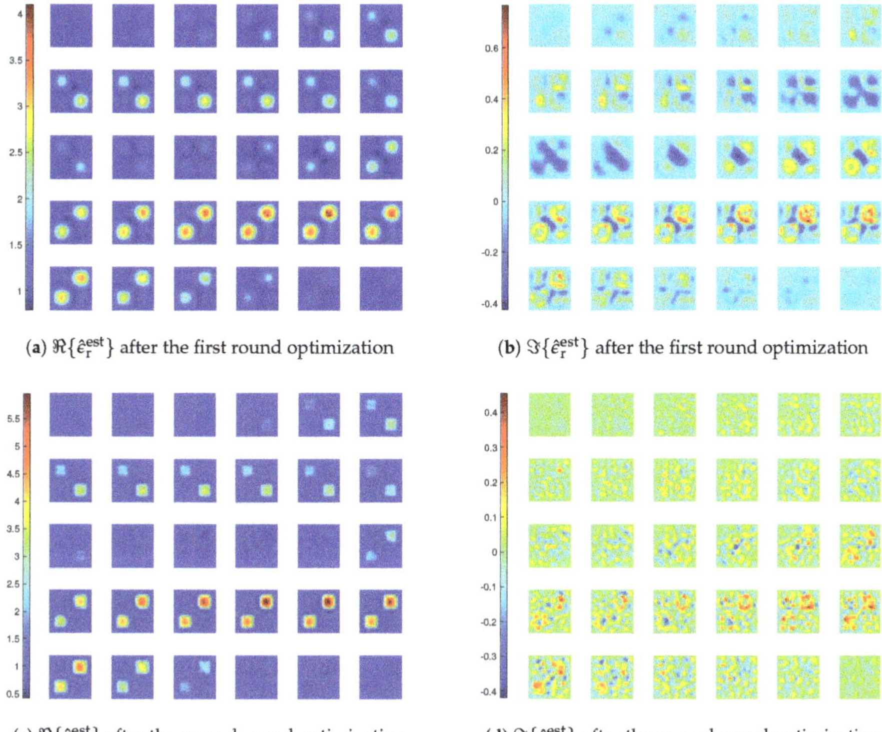

(**a**) $\Re\{\hat{\epsilon}_r^{est}\}$ after the first round optimization

(**b**) $\Im\{\hat{\epsilon}_r^{est}\}$ after the first round optimization

(**c**) $\Re\{\hat{\epsilon}_r^{est}\}$ after the second round optimization

(**d**) $\Im\{\hat{\epsilon}_r^{est}\}$ after the second round optimization

Figure 10. Reconstruction results of inversions with CIE-I for *Example 2*. In the first round, $\beta_1 = \beta_2 = 6$. In the second round, $\beta_1 = \beta_2 = 1$.

3.3. Example 3

We will use another highly nonlinear example to confirm the large difference between the performances of inversion methods with CIE-I and LSIE. In this example, we use a profile that is similar to the famous two-dimensional "Austria" that has an annular and two separated disks. Here, we use a coated cube with a hollow inside and two rods outside the cube, as shown in Figure 11, all of which are with $\epsilon_r = 2.5$.

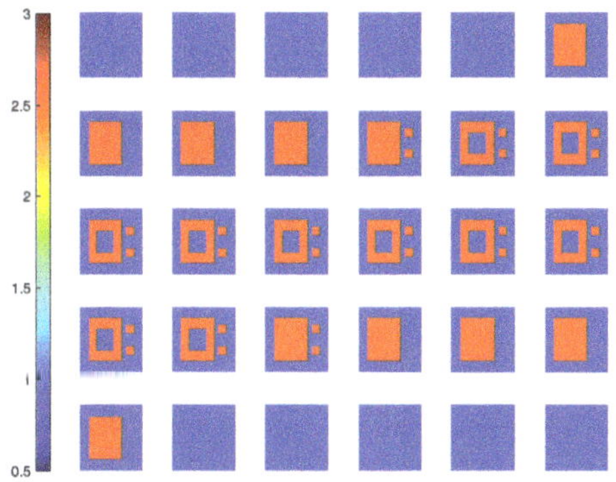

Figure 11. True permittivity profile (real part) for *Example 3*.

We still carry out inversions with LSIE and CIE-I with two rounds of optimization, the first with $M_F = 6$ and the second with $M_F = 10$. For the CIE-I, $\beta_1 = \beta_2 = 6$ is used in the first round, and $\beta_1 = \beta_2 = 1$ in the second round. The reconstructions and errors are shown in Figures 12, 13a and 14. We see that the inversion with LSIE fails to find the profile while the one with CIE-I succeeds. This can also be seen from the errors of the reconstructions in Figure 13a, where the error by the inversion with LSIE diverges.

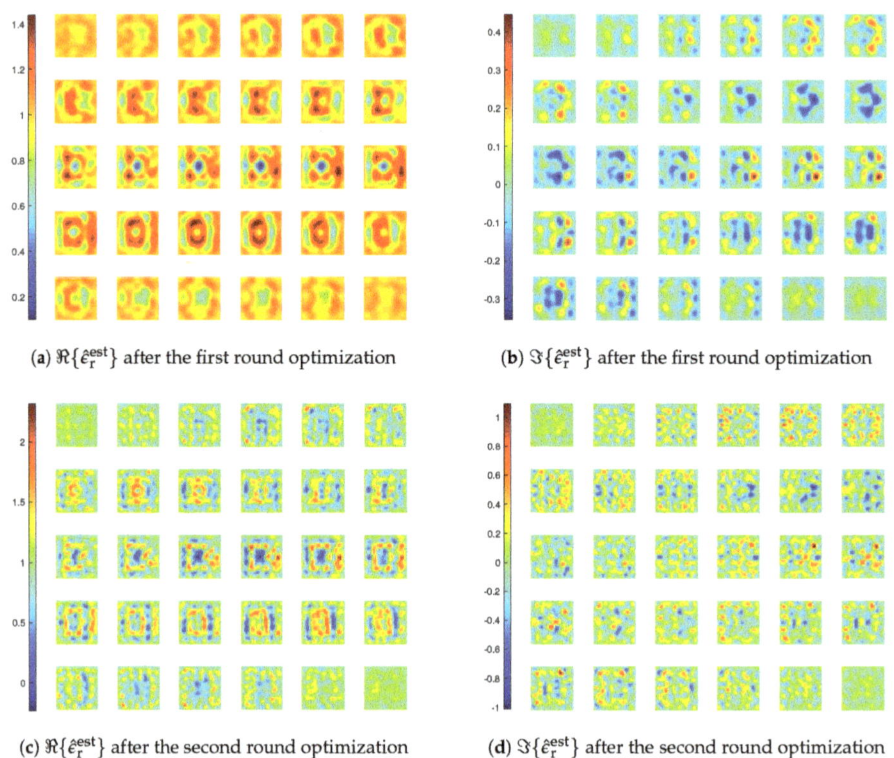

(**a**) $\Re\{\hat{\epsilon}_r^{\text{est}}\}$ after the first round optimization

(**b**) $\Im\{\hat{\epsilon}_r^{\text{est}}\}$ after the first round optimization

(**c**) $\Re\{\hat{\epsilon}_r^{\text{est}}\}$ after the second round optimization

(**d**) $\Im\{\hat{\epsilon}_r^{\text{est}}\}$ after the second round optimization

Figure 12. Reconstruction results of inversions with LSIE for *Example 3*.

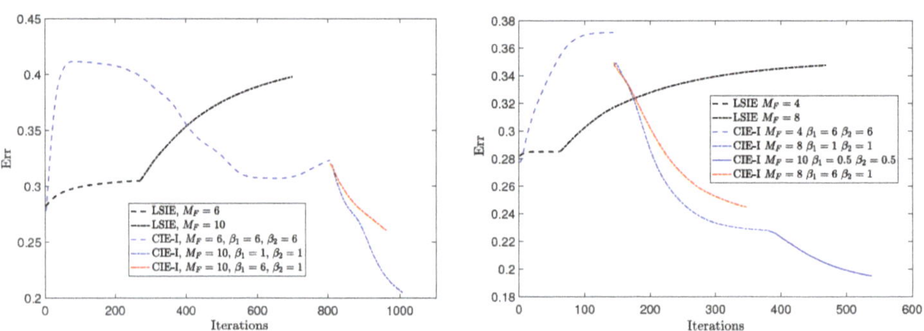

(**a**) Errors of reconstructions in inversions with $M_F = 6$ and 10.

(**b**) Errors of reconstructions in inversions with $M_F = 4$, 8, and 10.

Figure 13. Errors of reconstructions obtained by different inversion methods for *Example 3*.

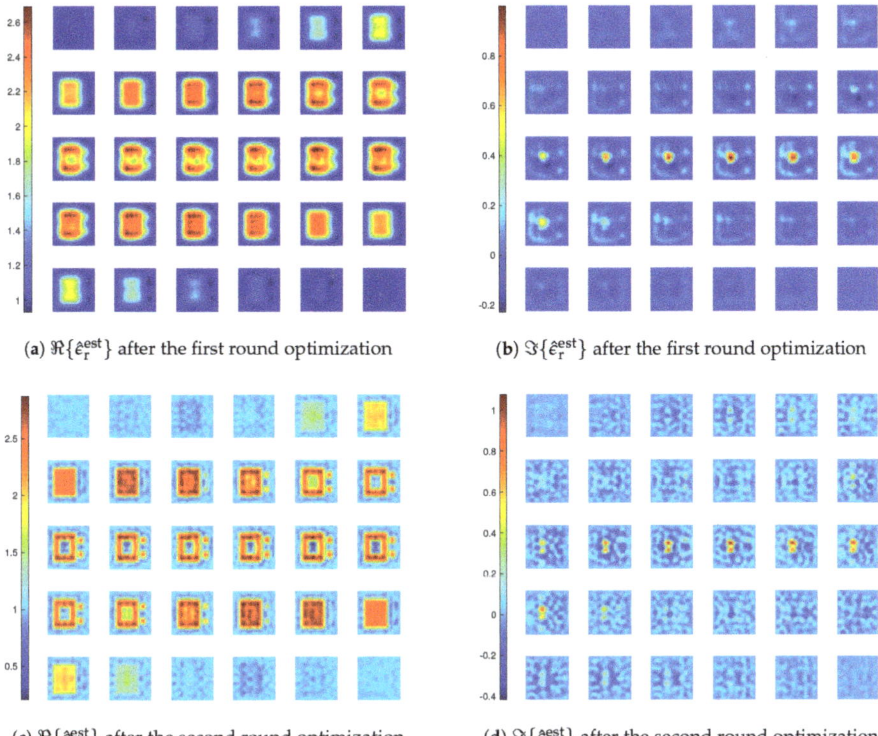

(**a**) $\Re\{\hat{\epsilon}_r^{est}\}$ after the first round optimization

(**b**) $\Im\{\hat{\epsilon}_r^{est}\}$ after the first round optimization

(**c**) $\Re\{\hat{\epsilon}_r^{est}\}$ after the second round optimization

(**d**) $\Im\{\hat{\epsilon}_r^{est}\}$ after the second round optimization

Figure 14. Reconstruction results of inversions with CIE-I for *Example 3*. In the first round, $\beta_1 = \beta_2 = 6$. In the second round, $\beta_1 = \beta_2 = 1$.

In the reconstruction results in Figure 14, we see that there is an artefact within the coated cube, where the medium is supposed to be air. This artefact appears in the reconstructed result after the first round optimization with $M_F = 6$, meaning that the inversion, though capable of retrieving most of the main features of the scatterers, is still not stable enough. Therefore, we carry out another inversion with three rounds of optimization, with $M_F = 4, 8$, and 10 in each round, and $\beta = 6, 1$, and 0.5, respectively. The final reconstruction result is given in Figure 15, and we see that there is no more such an artefact.

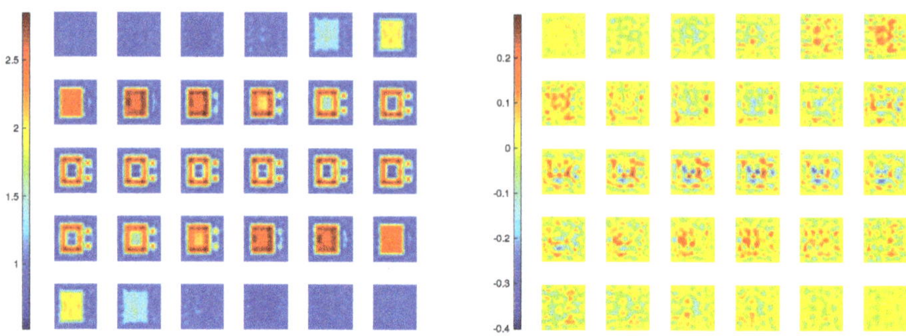

(a) $\Re\{\hat{\epsilon}_r^{est}\}$ after three rounds of optimization   (b) $\Im\{\hat{\epsilon}_r^{est}\}$ after three rounds of optimization

Figure 15. Reconstruction results of inversions with CIE-I after three rounds of optimization for *Example 3*. In the first round, $\beta_1 = \beta_2 = 6$ and $M_F = 4$. In the second round, $\beta_1 = \beta_2 = 1$ and $M_F = 8$. In the third round, $\beta_1 = \beta_2 = 0.5$ and $M_F = 10$.

At last, we also carry out the inversions with different β values for the updates of the APCS and the contrast, and the results are shown in Figure 13. We see that with larger β_1 value for the update of the APCS in the second round optimization, the convergence speed is indeed faster.

From this example, we again confirm the great resolvability of the inversion method with CIE-I against the high nonlinearity of the ISPs.

4. Conclusions

In this paper, we propose an inversion method with CIE-I to tackle the 3-D electromagnetic ISPs, especially for highly nonlinear problems with strong scatterers, those with high contrasts or electrically large dimensions. With the CIE-I modeling, the multiple scattering effects in estimating the contrast can be suppressed such that the nonlinearity of inversion can be effectively alleviated. Together with the FFT-TSOM regularization scheme, this largely increases the ability of retrieving the profile of strong scatterers, and effectively accelerates the convergence when handling the moderate scatterers. In addition, by relaxing the contraction mapping, i.e., different contraction mappings are used in updating the contrast sources and in updating the contrast, the convergence of the inversions can be further accelerated. Through numerical examples, we clearly show that the inversions with CIE-I outperform the ones with LSIE, in terms of the resolvability against the nonlinearity and the convergence speed. As shown in [32], the resolvability of the inversion solver with CIE-I against nonlinearity can be further improved in the two-dimensional case if proper regularization techniques are implemented, which should be expected in the 3-D case as well.

Author Contributions: Conceptualization, Y.Z.; Methodology, Y.Z.; Software, Y.Z. and K.X.; Validation, Y.Z., and K.X.; Formal Analysis, Y.Z. and K.X.; Investigation, Y.Z.; Resources, K.X.; Data Curation, Y.Z. and K.X.; Writing—Original Draft Preparation, Y.Z. and K.X.; Writing—Review and Editing, Y.Z., and K.X.; Visualization, Y.Z.; Supervision, Y.Z.

Funding: K.X. was partially funded by NSFC Grant No. 61601161.

Acknowledgments: The authors would like to thank Dominique Lesselier at Laboratoire des Signaux et Systèmes (UMR8506 CNRS-CentraleSupélec-Université Paris-Sud), France, for his very kind help in discussions of 3-D inversion methods and in writing the manuscript.

Conflicts of Interest: The authors declare no conflict of interest.

Abbreviations

The following abbreviations are used in this manuscript:

CIE-I   Contraction integral equation for inversion
LSIE   Lippmann–Schwinger integral equation
3-D   Three-dimensional
CSI   Contrast source inversion
SOM   Subspace-based optimization method
FFT-TSOM   FFT type twofold subspace-based optimization method
APCS   Ambiguous part of the contrast source
DPCS   Deterministic part of the contrast source
DoI   Domain of interest
MSE   Multiple scattering effects

References

1. Colton, D.; Kress, R. *Inverse Acoustic and Electromagnetic Scattering Theory*; Springer: New York, NY, USA, 2013.
2. Abubakar, A.; van den Berg, P.M.; Mallorqui, J. Imaging of biomedical data using a multiplicative regularized contrast source inversion method. *IEEE Trans. Microw. Theory Tech.* **2002**, *50*, 1761–1771. [CrossRef]
3. Abubakar, A.; van den Berg, P.M. Three-dimensional inverse scattering applied to cross-well induction sensors. *IEEE Trans. Antennas Propag.* **2000**, *38*, 1669–1681. [CrossRef]
4. Massa, A.; Boni, A.; Donelli, M. A classification approach based on SVM for electromagnetic subsurface sensing. *IEEE Trans. Antennas Propag.* **2005**, *43*, 2084–2093. [CrossRef]
5. Sabatier, P.C. Past and future of inverse problems. *J. Math. Phys.* **2000**, *41*, 4082–4124. [CrossRef]
6. Abubakar, A.; van den Berg, P.M. Iterative forward and inverse algorithms based on domain integral equations for three-dimensional electric and magnetic objects. *J. Comput. Phys.* **2004**, *195*, 236–262. [CrossRef]
7. Zhong, Y.; Chen, X.; Agarwal, K. An improved subspace-based optimization method and its implementation in solving three-dimensional inverse problems. *IEEE Trans. Geosci. Remote Sens.* **2010**, *48*, 3763–3768. [CrossRef]
8. Zhong, Y.; Chen, X. An FFT twofold subspace-based optimization method for solving electromagnetic inverse scattering problems. *IEEE Trans. Antennas Propag.* **2011**, *59*, 914–927. [CrossRef]
9. Litman, A.; Lorenzo, C. Special section on testing inversion algorithms against experimental data: 3-D targets. *Inverse Probl.* **2009**, *25*, 020201. [CrossRef]
10. Van den Berg, P.M.; Kleinman, R.E. A contrast source inversion method. *Inverse Probl.* **1997**, *13*, 1607–1620. [CrossRef]
11. Van den Berg, P.M.; van Broekhoven, A.L.; Abubakar, A. Extended constrast source inversion. *Inverse Probl.* **1999**, *15*, 1325–1344. [CrossRef]
12. Wang, Y.; Chew, W.C. An iterative solution of two-dimensional electromagnetic inverse scattering problem. *Int. J. Imaging Syst. Technol.* **1989**, *1*, 100–108. [CrossRef]
13. Chew, W.C.; Wang, Y. Reconstruction of two-dimensional permittivity distribution using the distorted Born iterative method. *IEEE Trans. Med. Imaging* **1990**, *9*, 218–225. [CrossRef] [PubMed]
14. Dorn, O.; Lesselier, D. Level set methods for inverse scattering. *Inverse Probl.* **2006**, *22*, R67–R131. [CrossRef]
15. Benedetti, M.; Lesselier, D.; Lambert, M.; Massa, A. A multi-resolution technique based on shape optimization for the reconstruction of homogeneous dielectric objects. *Inverse Probl.* **2009**, *25*, 015009. [CrossRef]
16. Chen, X. Subspace-based optimization method for solving inverse scattering problems. *IEEE Trans. Geosci. Remote Sens.* **2010**, *48*, 42–49. [CrossRef]
17. Zhong, Y.; Chen, X. Twofold subspace-based optimization method for solving inverse scattering problems. *Inverse Probl.* **2009**, *25*, 085003. [CrossRef]
18. Agarwal, K.; Pan, L.; Chen, X. Subspace-based optimization method for reconstruction of two-dimensional complex anisotropic dielectric objects. *IEEE Trans. Microw. Theory Tech.* **2009**, *58*, 1065–1074. [CrossRef]
19. Pastorino, M. Stochastic optimization methods applied to microwave imaging: A review. *IEEE Trans. Antenna Propag.* **2007**, *55*, 538–548. [CrossRef]
20. Rocca, P.; Benedetti, M.; Donelli, M.; Franceschini, D.; Massa, A. Evolutionary optimization as applied to inverse scattering problems. *Inverse Probl.* **2009**, *25*, 123003. [CrossRef]

21. De Zaeytijd, J.; Franchois, A.; Geffrin, J.M. A new value picking regularization strategy-Application to the 3-D electromagnetic inverse scattering problem. *IEEE Trans. Antenna Propag.* **2009**, *57*, 1133–1149. [CrossRef]
22. Chaumet, P.; Belkebir, K. Three-dimensional reconstruction from real data using a conjugate gradient-coupled dipole method. *Inverse Probl.* **2009**, *25*, 024003. [CrossRef]
23. Yu, C.; Yuan, M.; Liu, Q.H. Reconstruction of 3-D objects from multi-frequency experimental data with a fast DBIM-BCGS method. *Inverse Probl.* **2009**, *25*, 024007. [CrossRef]
24. Donelli, M.; Franceschini, D.; Rocca, P.; Massa, A. Three-dimensional microwave imaging problems solved through an efficient multiscaling particle swarm optimization. *IEEE Trans. Geosci. Remote Sens.* **2009**, *47*, 1467–1481. [CrossRef]
25. Agarwal, K.; Chen, X.; Zhong, Y. A multipole-expansion based linear sampling method for solving inverse scattering problems. *Opt. Express* **2010**, *18*, 6366–6381. [CrossRef] [PubMed]
26. Bevacqua, M.T.; Isernia, T. Boundary Indicator for Aspect Limited Sensing of Hidden Dielectric Objects. *IEEE Geosci. Remote Sens. Lett.* **2018**, *15*, 838–842. [CrossRef]
27. Isernia, T.; Crocco L.; D'Urso M. New tools and series for forward and inverse scattering problems in lossy media. *IEEE Geosci. Remote Sens. Lett.* **2004**, *1*, 327–331. [CrossRef]
28. D'Urso, M.; Isernia T.; Morabito, A.F. On the Solution of 2-D Inverse Scattering Problems via Source-Type Integral Equations. *IEEE Trans. Geosci. Remote Sens.* **2010**, *48*, 1186–1198. [CrossRef]
29. Zhong, Y.; Lambert, M.; Lesselier, D.; Chen, X. A new integral equation method to solve highly nonlinear inverse scattering problems. *IEEE Trans. Antennas Propag.* **2016**, *64*, 1788–1799. [CrossRef]
30. Pankratov, O.V.; Avdeyev, D.B.; Kuvshinov, A.V. Electromagnetic field scattering in a heterogeneous earth: A solution to the forward problem. *Phys. Solid Earth* **1995**, *31*, 201–209.
31. Peterson, A.F.; Ray, S.L.; Mittra, R. *Computational Methods for Electromagnetics*; IEEE Press: New York, NY, USA, 1998.
32. Xu, K.; Zhong, Y.; Wang, G. A hybrid regularization technique for solving highly nonlinear inverse scattering problems. *IEEE Trans. Microw. Theory Tech.* **2018**, *64*, 11–21. [CrossRef]

Journal of
Imaging

Article

Full-Vectorial *3D* Microwave Imaging of Sparse Scatterers through a Multi-Task Bayesian Compressive Sensing Approach

Marco Salucci [1,2,†], **Lorenzo Poli** [1,2,†] **and Giacomo Oliveri** [1,2,*,†]

1   ELEDIA Research Center (ELEDIA@UniTN—University of Trento), Via Sommarive 9, I-38123 Trento, Italy; marco.salucci@unitn.it (M.S.); lorenzo.poli@unitn.it (L.P.)
2   ELEDIA Research Center (ELEDIA@L2S—UMR 8506), 3 rue Joliot Curie, 91192 Gif-sur-Yvette, France
*   Correspondence: giacomo.oliveri@unitn.it
†   These authors contributed equally to this work.

Received: 3 December 2018; Accepted: 8 January 2019; Published: 15 January 2019

Abstract: In this paper, the full-vectorial three-dimensional (*3D*) microwave imaging (*MI*) of sparse scatterers is dealt with. Towards this end, the inverse scattering (*IS*) problem is formulated within the contrast source inversion (*CSI*) framework and it is aimed at retrieving the sparsest and most probable distribution of the contrast source within the imaged volume. A customized multi-task Bayesian compressive sensing (*MT-BCS*) method is used to yield regularized solutions of the *3D-IS* problem with a remarkable computational efficiency. Selected numerical results on representative benchmarks are presented and discussed to assess the effectiveness and the reliability of the proposed *MT-BCS* strategy in comparison with other competitive state-of-the-art approaches, as well.

Keywords: microwave imaging; inverse scattering; Bayesian compressive sensing (*BCS*); contrast source inversion (*CSI*); 3D

1. Introduction

Microwave imaging (*MI*) techniques are aimed at inferring the complex permittivity distribution within an inaccessible investigation domain from the scattering interactions between the matter and probing electromagnetic (*EM*) waves [1]. They have been successfully applied in several diagnostic scenarios including non-destructive testing and evaluation [2,3], through-wall imaging [4], subsurface prospecting [5–10], and structural health monitoring [11]. Moreover, they represent a very appealing technology in many biomedical applications [12] such as, for instance, breast cancer detection [13–22] thanks to the use of non-ionizing radiations. To date, significant efforts have been mostly devoted to the development of two-dimensional (*2D*) *MI* algorithms, mainly based on transverse-magnetic (*TM*) [23–28] or transverse-electric (*TE*) polarized [29] configurations, rather than fully-vectorial three-dimensional (3D) ones [30]. As a matter of fact, under the assumption that the *EM* properties of the unknown scattering scenario are invariant along a longitudinal direction, the arising inverse scattering (*IS*) problem can be recast to the solution of simplified scalar Helmholtz equations [1]. On the other hand, 2D-MI approaches are prone to errors when finite-volume scatterers are under test [31] because of the over-simplified modelling of the scattering phenomena. The slower evolution of 3D-MI techniques has been mainly caused by the higher complexity of both data-collection/storage and image reconstruction processes with respect to the tomographic (2D) case. Moreover, a significantly larger number of unknowns has to be retrieved and it becomes very hard to manage when there is the need of high-resolution images (e.g., a realistic discretization of a human thorax needs millions of voxels for having a clinical significance [32]). Furthermore, the amount of non-redundant information on the investigation domain achievable from measurements is upper-bounded and the ratio between unknowns and scattering data turns out to be very high [33]. Owing to such limitations, solving *3D-MI* problems faces hard challenges and it requires the non-trivial implementation of effective countermeasures to both the *non-linearity* and the *ill-posedness* issues of the arising full-vectorial *IS* problem.

Dealing with *3D* scenarios, synthetic aperture radar (*SAR*)-based methodologies such as confocal *MI* [34] and synthetic near-field focusing [5] have been proposed. They are based on the emission of wide-band pulses from multiple transmitting positions and the successive processing of the collected echoes. Only target detection and localization (i.e., a *qualitative* imaging of the investigation domain) is typically yielded, while *quantitatively* retrieving the distribution of the *EM* properties needs the numerical solution of the non-linear scattering equations. Towards this end, effective *3D-MI* approaches have been presented in the scientific literature. They are based on the processing of time-domain [19,35] or frequency-domain [36,37] data and, due to the extremely-wide dimension of the solution space, which is usually proportional to the number of discretization domains, state-of-the-art methodologies are primarily based on deterministic methods (e.g., Gauss-Newton (*GN*) [36,38], Conjugate-Gradient (*CG*) [39], Level Set (*LS*) [16], and Inexact-Newton (*IN*) [40] methods, possibly formulated in different functional spaces [41]), even though they do not *a priori* guarantee to reach the actual solution (i.e., the global optimum of the cost function quantifying the mismatch between measured and estimated scattering data). To overcome such a drawback, either very efficient forward solvers [42,43] have been introduced or stochastic multi-agent inversion algorithms, suitably integrated with multi-resolution strategies [37] for a sustainable customization to the *3D* case, have been successfully exploited. Alternatively, computationally-efficient approaches to the *MI* problem have been recently explored within the compressive sensing (*CS*) framework [23,29,44–49]. Despite the early stage [47,50] of their implementation in *3D* cases, very interesting applicative examples are already available [20].

According to the *CS* theory, *sparseness priors* can be enforced to solve the *IS* problem and to yield a regularized solution provided that it admits a representation with few non-null coefficients in a suitably chosen basis [44,45]. However, since available *CS* solvers generally deal with linear problems, many sparsity-promoting approaches have been formulated within Born-like approximations, their success being limited to weak scatterers or to specific applications where a qualitative imaging is enough [29]. Alternatively, contrast source inversion (*CSI*)-based formulations of the *IS* problem can be successfully employed to yield accurate reconstructions also in the presence of scatterers with high *EM* contrast with respect to the surrounding medium [23]. Following such a line of reasoning, this paper is aimed at presenting a novel computationally-efficient approach, based on a Bayesian *CS* (*BCS*) method, to solve the *3D-IS* problem concerned with non-weak scatterers. Towards this end, the full-vectorial *IS* problem is formulated within a probabilistic *CSI* framework and then it is efficiently solved through a customized multi-task *BCS* (*MT-BCS*) strategy.

The outline of the paper is as follows. The formulation of the *3D-CSI MI* problem is detailed in Section 2, while the proposed *BCS*-based solution strategy is described in Section 3. Selected numerical results, from representative test cases, are presented and compared with competitive state-of-the-art alternatives in Section 4. Finally, some concluding remarks are drawn (Section 5).

2. Mathematical Formulation

Let us consider a *3D* isotropic non-magnetic [$\mu(\mathbf{r}) = \mu_0$] scattering scenario characterized by a relative permittivity distribution, $\varepsilon(\mathbf{r})$, and a conductivity profile, $\sigma(\mathbf{r})$, $\mathbf{r}$ being the position vector defined as $\mathbf{r} = x\mathbf{u}_x + y\mathbf{u}_y + z\mathbf{u}_z$, $\mathbf{u}_p$ being the unit vector along the *p*-th ($p = \{x, y, z\}$) direction. The goal of the *MI* problem at hand is to estimate the contrast function [51]

$$\tau(\mathbf{r}) \triangleq [\varepsilon(\mathbf{r}) - 1] - j\frac{\sigma(\mathbf{r}) - \sigma_0}{\omega\varepsilon_0} \tag{1}$$

within the investigation domain $\mathcal{D}$ (i.e., $\forall \mathbf{r} \in \mathcal{D}$), $\omega = 2\pi f$, ε_0, and σ_0 being the angular frequency (A time dependency factor $\exp(j\omega t)$ is assumed and omitted hereinafter to simplify the notation, but without loss of generality in the mathematical formulation.), the background permittivity and conductivity (with $\sigma_0 = 0$ hereinafter), respectively, and $\triangleq$ standing for "defined as". Towards this end, a set of V monochromatic (f being the working frequency) plane-waves impinging from known angular directions (θ_v, φ_v), $v = 1, ..., V$, with known electric field

$$\mathbf{E}_v^i(\mathbf{r}) = \sum_{p=\{x,y,z\}} E_{v,p}^i(\mathbf{r})\,\mathbf{u}_p \quad v = 1, ..., V \tag{2}$$

is used to successively probe the unaccessible domain $\mathcal{D}$ (Figure 1).

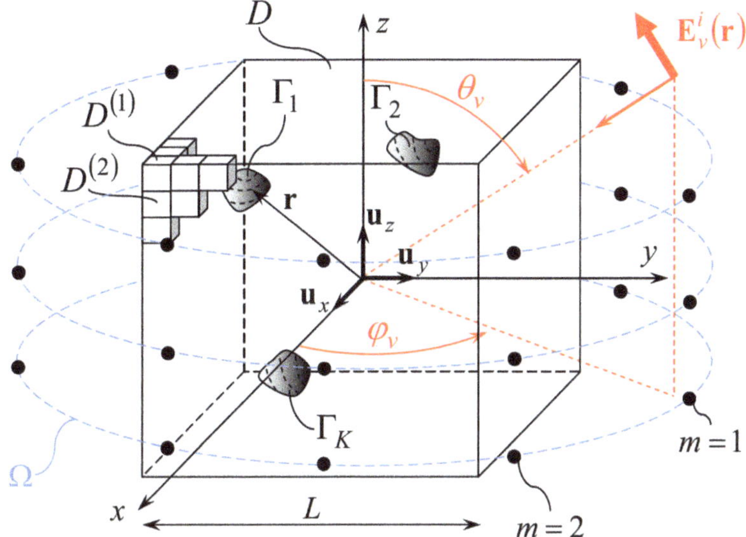

Figure 1. Geometry of the 3D-MI microwave imaging problem.

Under such hypotheses and by adopting a *CSI* formulation [52] for the scattering problem, the electromagnetic interactions between the scatterers in $\mathcal{D}$ and the v-th ($v = 1, ..., V$) incident wave can be mathematically described through the following integral *data* equation

$$\mathbf{E}_v^s(\mathbf{r}) = -\omega^2 \varepsilon_0 \mu_0 \int \int \int_{\mathcal{D}} \mathbf{J}_v(\mathbf{r}') \cdot \mathbf{G}(\mathbf{r}, \mathbf{r}')\,d\mathbf{r}' \quad \mathbf{r} \in \Omega \tag{3}$$

where Ω is an observation domain external to $\mathcal{D}$ ($\Omega \cap \mathcal{D} = \varnothing$—Figure 1), $\mathbf{E}_v^s(\mathbf{r})$ [$\mathbf{E}_v^s(\mathbf{r}) = \sum_{p=\{x,y,z\}} E_{v,p}^s(\mathbf{r})\,\mathbf{u}_p$] is the v-th ($v = 1, ..., V$) *scattered* field defined as the difference between the v-th ($v = 1, ..., V$) electric field with, $\mathbf{E}_v(\mathbf{r}) = \sum_{p=\{x,y,z\}} E_{v,p}(\mathbf{r})\,\mathbf{u}_p$ (i.e., the v-th *total* electric field), and without, $\mathbf{E}_v^i(\mathbf{r})$ (i.e., the v-th *incident* electric field), the scatterers in the background medium ($\mathbf{E}_v^s(\mathbf{r}) \triangleq [\mathbf{E}_v(\mathbf{r}) - \mathbf{E}_v^i(\mathbf{r})]$), $\cdot$ stands for the scalar product, and

$$\mathbf{G}(\mathbf{r}, \mathbf{r}') = \sum_{q=\{x,y,z\}} \sum_{p=\{x,y,z\}} G_{pq}\mathbf{u}_p\mathbf{u}_q = \frac{1}{4\pi}\left(\mathbf{I} + \frac{1}{\omega^2 \varepsilon_0 \mu_0}\nabla\nabla\right)\frac{\exp\left(-j\omega\sqrt{\varepsilon_0\mu_0}\,|\mathbf{r}-\mathbf{r}'|\right)}{|\mathbf{r}-\mathbf{r}'|} \tag{4}$$

Is the dyadic Green's function for the homogeneous free-space background medium of dielectric and magnetic properties ε_0 and μ_0, respectively, $\mathbf{I}$ being the unit tensor. In (3), $\mathbf{J}_v$ [$\mathbf{J}_v(\mathbf{r}) =$

$\sum_{p=\{x,y,z\}} J_{v,p}(\mathbf{r}) \mathbf{u}_p]$ is the v-th ($v = 1, ..., V$) unknown contrast current density induced in the investigation domain ($\mathbf{r} \in \mathcal{D}$) by the v-th probing field (i.e., the v-th illumination of $\mathcal{D}$)

$$\mathbf{J}_v(\mathbf{r}) = \tau(\mathbf{r}) \sum_{p=\{x,y,z\}} E_{v,p}(\mathbf{r}) \mathbf{u}_p \tag{5}$$

that models the scattering profile of $\mathcal{D}$ and it is defined as the v-th ($v = 1, ..., V$) *equivalent* source radiating in the background medium an electromagnetic field equal to the v-th ($v = 1, ..., V$) *scattered* field $\mathbf{E}_v^s(\mathbf{r})$. Furthermore, the v-th ($v = 1, ..., V$) incident field $\mathbf{E}_v^i(\mathbf{r})$ complies with the so-called integral *state* equation

$$\mathbf{E}_v^i(\mathbf{r}) = \mathbf{E}_v(\mathbf{r}) + \omega^2 \varepsilon_0 \mu_0 \int \int \int_{\mathcal{D}} \mathbf{J}_v(\mathbf{r}') \cdot \mathbf{G}(\mathbf{r}, \mathbf{r}')\, d\mathbf{r}' \tag{6}$$

within the investigation domain ($\mathbf{r} \in \mathcal{D}$).

To numerically deal with the inverse problem at hand, a set of N 3D rectangular pulse functions

$$\Psi^{(n)}(\mathbf{r}) = \begin{cases} 1 & \text{if } \mathbf{r} \in \mathcal{D}^{(n)} \\ 0 & \text{otherwise} \end{cases} \quad ; \quad n = 1, ..., N \tag{7}$$

defined in a set of N cubic (also called *voxels*) sub-domains ($\mathcal{D} = \cup_{n=1}^{N} \mathcal{D}^{(n)}$—Figure 1) is adopted for yielding the following piece-wise constant representation of the v-th ($v = 1, ..., V$) unknown contrast source

$$\mathbf{J}_v(\mathbf{r}) = \sum_{p=\{x,y,z\}} \sum_{n=1}^{N} J_{v,p}^{(n)} \Psi^{(n)}(\mathbf{r}) \mathbf{u}_p \tag{8}$$

where $J_{v,p}^{(n)} = J_{v,p}(\mathbf{r}_n)$ ($p = \{x, y, z\}$) and $\mathbf{r}_n$ denotes the barycentre of the n-th ($n = 1, ..., N$) voxel $\mathcal{D}^{(n)}$.

By sampling the scattered field in M probing locations of Ω ($\mathbf{r}_m \in \Omega$, $m = 1, ..., M$), it is possible to rewrite (3) in the following discrete form [1]

$$E_v^s(\mathbf{r}_m) = \sum_{p=\{x,y,z\}} \sum_{q=\{x,y,z\}} \sum_{n=1}^{N} J_{v,q}^{(n)} G_{pq}^{(mn)} \mathbf{u}_p, \tag{9}$$

where $G_{pq}^{(mn)} = -\omega^2 \varepsilon_0 \mu_0 \int \int \int_{\mathcal{D}^{(n)}} G_{pq}(\mathbf{r}_m, \mathbf{r}')\, d\mathbf{r}'$ ($p, q = \{x, y, z\}$), or in compact matrix notation as follows

$$\left[\underline{E}_{v,x}^s, \underline{E}_{v,y}^s, \underline{E}_{v,z}^s \right]^T = \underline{\underline{G}}^{ext} \left[\underline{J}_{v,x'}, \underline{J}_{v,y'}, \underline{J}_{v,z} \right]^T \tag{10}$$

where $.^T$ indicates the transpose operator, $\underline{E}_{v,p}^s \triangleq \left[E_{v,p}^s(\mathbf{r}_m) ; m = 1, ..., M \right]$ ($\underline{E}_{v,p}^s \in \mathbb{C}^{1 \times M}$), $\underline{J}_{v,p} \triangleq \left[J_{v,p}^{(n)} ; n = 1, ..., N \right]$ ($\underline{J}_{v,p} \in \mathbb{C}^{1 \times N}$), $p = \{x, y, z\}$, and

$$\underline{\underline{G}}^{ext} \triangleq \begin{bmatrix} \underline{\underline{G}}_{xx}^{ext} & \underline{\underline{G}}_{xy}^{ext} & \underline{\underline{G}}_{xz}^{ext} \\ \underline{\underline{G}}_{yx}^{ext} & \underline{\underline{G}}_{yy}^{ext} & \underline{\underline{G}}_{yz}^{ext} \\ \underline{\underline{G}}_{zx}^{ext} & \underline{\underline{G}}_{zy}^{ext} & \underline{\underline{G}}_{zz}^{ext} \end{bmatrix} \tag{11}$$

is the external 3D Green's operator ($\underline{\underline{G}}^{ext} \in \mathbb{C}^{3M \times 3N}$) where the (m, n)-th ($m = 1, ..., M; n = 1, ..., N$) entry of the (p, q)-th ($p, q = \{x, y, z\}$) sub-matrix, $\underline{\underline{G}}_{pq}^{ext} \in \mathbb{C}^{M \times N}$, is equal to $G_{pq}^{(mn)}$.

Under the hypothesis that the contrast distribution in $\mathcal{D}$ is *sparse* with respect to the basis at hand (7), it also turns out that the v-th ($v = 1, ..., V$) contrast source (8) can be numerically modeled with just $3 \times O$ ($O \ll N$) non-zero coefficients $\left\{ \left[J_{v,x}^{(o)}; J_{v,y}^{(o)}; J_{v,z}^{(o)} \right] ; o = 1, ..., O \right\}$, O being the number of sub-domains $\mathcal{D}^{(o)} \in \mathcal{D}$, $o = 1, ..., O$, occupied by scatterers (i.e., dielectric discontinuities of the background medium within the investigation domain).

3. Inversion Method

The matrix relation in (10) is representative of a severely *ill-posed* problem being (*i*) *ill-defined* since its solution is not-unique due to the presence of non-radiating sources in $\mathcal{D}$ that afford a null/not-measurable field in Ω and (*ii*) *ill-conditioned* since the condition number of $\underline{\underline{G}}^{ext}$, η, is large ($\eta \gg 1$). To counteract such a negative feature for yielding stable reconstructions of the *EM* properties of the investigation domain from the solution of (10) when processing noisy data, as well, enforcing the *a-priori* knowledge of the *sparseness* of the unknown equivalent source with respect to a suitable basis turns out to be an effective regularization strategy. Towards this end, the *CS* formulation based on the Bayesian theory is adopted to re-formulate the *3D-CSI* problem at hand in a *probabilistic* sense. Such a choice is done to avoid the need of numerically checking the fulfillment of the restricted isometry property (*RIP*) of the 3D Green's operator (11) as required by deterministic *CS* solvers. Indeed, whether such a compliance test is already very hard from a computational viewpoint for small-scale problems, it becomes computationally unfeasible in *3D* scattering scenarios since exponentially heavier than for *2D* cases.

In order to apply computationally-efficient *BCS* solvers, (10) is first rewritten as a larger real-valued linear problem

$$\underline{\mathcal{E}}_{v,p} = \underline{\underline{\mathcal{G}}}_p \underline{\mathcal{J}}_v \tag{12}$$

where $\underline{\mathcal{E}}_{v,p} \triangleq \left[\mathcal{R}\left(\underline{E}^s_{v,p} \right), \mathcal{I}\left(\underline{E}^s_{v,p} \right) \right]^T$ and $\underline{\mathcal{J}}_v \triangleq \left[\mathcal{R}\left(\underline{J}_{v,p};\, p = \{x, y, z\} \right), \mathcal{I}\left(\underline{J}_{v,p};\, p = \{x, y, z\} \right) \right]^T$ are the *p*-th ($p = \{x, y, z\}$) data vector, $\underline{\mathcal{E}}_{v,p} \in \mathbb{R}^{2M \times 1}$, and the *v*-th ($v = 1, ..., V$) unknown source vector, $\underline{\mathcal{J}}_v \in \mathbb{R}^{6N \times 1}$, respectively, while the Green's matrix operator, $\underline{\underline{\mathcal{G}}}_p \in \mathbb{R}^{2M \times 6N}$, is given by

$$\underline{\underline{\mathcal{G}}}_p \triangleq \left[\begin{array}{cc} \mathcal{R}\left(\begin{array}{ccc} \underline{\underline{G}}^{ext}_{px} & \underline{\underline{G}}^{ext}_{py} & \underline{\underline{G}}^{ext}_{pz} \end{array} \right) & -\mathcal{I}\left(\begin{array}{ccc} \underline{\underline{G}}^{ext}_{px} & \underline{\underline{G}}^{ext}_{py} & \underline{\underline{G}}^{ext}_{pz} \end{array} \right) \\ \mathcal{I}\left(\begin{array}{ccc} \underline{\underline{G}}^{ext}_{px} & \underline{\underline{G}}^{ext}_{py} & \underline{\underline{G}}^{ext}_{pz} \end{array} \right) & \mathcal{R}\left(\begin{array}{ccc} \underline{\underline{G}}^{ext}_{px} & \underline{\underline{G}}^{ext}_{py} & \underline{\underline{G}}^{ext}_{pz} \end{array} \right) \end{array} \right] \tag{13}$$

$\mathcal{R}(.)$ and $\mathcal{I}(.)$ being the real and the imaginary part function, respectively. According to such a description, the original inverse scattering problem can be then formulated as follows

3D-CSI BCS-Based Problem Formulation—Starting from the measurement of $\underline{\mathcal{E}}_{v,p}$ ($v = 1, ..., V$, $p = \{x, y, z\}$), and the knowledge of $\underline{\underline{\mathcal{G}}}_p$ ($p = \{x, y, z\}$), determine the *sparsest* guess of $\underline{\mathcal{J}}_v$, $\underline{\hat{\mathcal{J}}}_v = \left\{ \hat{\mathcal{J}}_v^{(t)}; t = 1, ..., 6 \times N \right\}$ as the maximum *a-posteriori* probability (*MAP*) estimate

$$\underline{\hat{\mathcal{J}}}_v = \arg \left\{ \max_{\underline{\mathcal{J}}_v} \left[\mathcal{P}\left(\underline{\mathcal{J}}_v \middle| \underline{\mathcal{E}}_{v,p} \right) \right] \right\} \tag{14}$$

provided that the support of $\underline{\hat{\mathcal{J}}}_v$, $\mathcal{S}_v = \left\{ t \in [1, ..., 6 \times N] \middle|\, \hat{\mathcal{J}}_v^{(t)} \neq 0 \right\}$ ($v = 1, ..., V$) is the same for all V different illuminations (i.e., $\mathcal{S}_1 = \mathcal{S}_2 = ... = \mathcal{S}_V$).

In order to enforce the physical correlation existing among the V solutions of (14) as well as between the unknown contrast sources and the vectorial components of the known scattered field, a customized version of the *MT-BCS* solver [53] is adopted by setting the number of parallel "tasks" to $H = (3 \times V)$ (owing to the 3D nature of the problem). More in detail, (14) is firstly reformulated in the Bayesian *MT* framework as [29,53]

$$\underline{\hat{\mathcal{J}}}_v = \arg \left\{ \max_{\underline{\mathcal{J}}_v} \left[\frac{\mathcal{P}\left(\underline{\mathcal{E}}_{v,p} \middle| \underline{\mathcal{J}}_v \right) \mathcal{P}\left(\underline{\mathcal{J}}_v \right)}{\mathcal{P}\left(\underline{\mathcal{E}}_{v,p} \right)} \right] \right\} \tag{15}$$

starting from the definition of a shared prior $\mathcal{P}\left(\underline{\mathcal{J}}_v\right)$ that statistically links the $3 \times V$ *parallel* problem unknowns as [29,53]

$$\mathcal{P}\left(\underline{\mathcal{J}}_v\right) = \int \mathcal{P}\left(\underline{\mathcal{J}}_v \mid \underline{\psi}\right) d\underline{\psi} \tag{16}$$

where $\underline{\psi}$ ($\underline{\psi} \triangleq \left\{\psi^{(t)}; t = 1, ..., 6 \times N\right\}$) is the set of *shared*—among the V-views and the three Cartesian scattered field components—hyper-parameters. By assuming a hierarchical Gaussian model for $\mathcal{P}\left(\underline{\mathcal{J}}_v \mid \underline{\psi}\right)$ and substituting (16) in (15) [29,53], the MT-BCS v-th ($v = 1, ..., V$) source term is then given after simple manipulations [29] by the following expression

$$\underline{\hat{\mathcal{J}}}_v = \frac{1}{3} \sum_{p=\{x,y,z\}} \left(\underline{\underline{G}}_p^T \underline{\underline{G}}_p + \underline{\underline{A}}\right)^{-1} \underline{\underline{G}}_p^T \underline{\mathcal{E}}_{v,p} \tag{17}$$

where $\underline{\underline{A}} = diag\left(\underline{\hat{\psi}}\right)$ and $\underline{\hat{\psi}} \triangleq \left\{\hat{\psi}^{(t)}; t = 1, ..., 6 \times N\right\}$. These latter are determined with a computationally-efficient relevant vector machine (RVM) [54] by solving the following maximum marginal likelihood problem

$$\underline{\hat{\psi}} = \arg\left\{max_{\underline{\psi}}\left[\mathcal{L}\left(\underline{\psi} \mid \alpha, \beta\right)\right]\right\} \tag{18}$$

$\mathcal{L}\left(\underline{\psi} \mid \alpha, \beta\right)$ being the logarithmic marginal likelihood function for the fully-vectorial problem given by

$$\mathcal{L}\left(\underline{\psi} \mid \alpha, \beta\right) = -\tfrac{1}{2}\sum_{v=1}^{V}\sum_{p=\{x,y,z\}}\left\{2\left(M+\alpha\right)\log\left[\underline{\mathcal{E}}_{v,p}^T\right.\right.$$
$$\left.\left(\underline{\underline{I}} + \underline{\underline{G}}_p\underline{\underline{A}}^{-1}\underline{\underline{G}}_p^T\right)^{-1}\underline{\mathcal{E}}_{v,p} + 2\beta\right] + \log\left|\underline{\underline{I}} + \underline{\underline{G}}_p\underline{\underline{A}}^{-1}\underline{\underline{G}}_p^T\right|\right\} \tag{19}$$

where $\underline{\underline{I}}$ is the identity matrix, while α and β are user-defined control parameters. Once $\underline{\hat{\mathcal{J}}}_v$ ($v = 1, ..., V$) has been estimated through (17), the contrast function, $\hat{\tau}\left(\mathbf{r}\right) = \sum_{n=1}^{N}\hat{\tau}^{(n)}\Psi^{(n)}\left(\mathbf{r}\right)$, is retrieved by computing the corresponding expansion coefficient vector $\underline{\hat{\tau}} = \left\{\hat{\tau}^{(n)} \in \mathbb{C}; n = 1, ..., N\right\}$, whose n-th ($n = 1, ..., N$) entry is given by

$$\hat{\tau}^{(n)} = \sum_{p=\{x,y,z\}} \sum_{v=1}^{V} \frac{\hat{J}_{v,p}^{(n)}}{3 \times V \times \hat{E}_{v,p}\left(\mathbf{r}_n\right)} \tag{20}$$

where $\hat{E}_{v,p}\left(\mathbf{r}_n\right)$ is the p-th ($p = \{x, y, z\}$) component of the v-th ($v = 1, ..., V$) total field in the n-th voxel ($\mathbf{r}_n \in \mathcal{D}^{(n)}$) ($n = 1, ..., N$) yielded from the following field relation

$$\hat{E}_{v,p}\left(\mathbf{r}_n\right) = E_{v,p}^i\left(\mathbf{r}_n\right) - \omega^2\varepsilon_0\mu_0 \sum_{q=\{x,y,z\}}\sum_{n=1}^{N}\hat{J}_{v,q}^{(n)}\int\int\int_{\mathcal{D}^{(n)}}G_{pq}\left(\mathbf{r}_n, \mathbf{r}'\right)d\mathbf{r}', \tag{21}$$

while the coefficient $\hat{J}_{v,p}^{(n)}$ ($n = 1, ..., N; v = 1, ..., V; p = \{x, y, z\}$) is derived from $\underline{\hat{\mathcal{J}}}_v$ (17) according to the following mapping

$$\hat{J}_{v,p}^{(n)} = \begin{cases} \hat{\mathcal{J}}_v^{(n)} + j\hat{\mathcal{J}}_v^{(n+3\times N)} & \text{if } p = x \\ \hat{\mathcal{J}}_v^{(n+N)} + j\hat{\mathcal{J}}_v^{(n+4\times N)} & \text{if } p = y \\ \hat{\mathcal{J}}_v^{(n+2N)} + j\hat{\mathcal{J}}_v^{(n+5\times N)} & \text{if } p = z \end{cases} . \tag{22}$$

4. Numerical Assessment

In this Section, representative results from a wide numerical analysis are presented and discussed to assess the reconstruction capabilities and the robustness of the proposed 3D-MI approach. Moreover, some practical guidelines for the optimal setting of its key calibration parameters will be provided for helping the interested users. Towards this end, the following reference scenario has been considered.

A cubic volume of side $L = 1.25\ [\lambda]$, λ being the free-space wavelength, has been chosen as investigation domain $\mathcal{D}$ and it has been probed by $V = 48$ plane-waves impinging from the V angular directions $(\theta_v, \varphi_v) = \left(\frac{\pi}{2}, (v-1)\frac{2\pi}{V}\right)$, $v = 1, ..., V$. The scattered field has been collected in $M = 48$ locations

$$\begin{cases} x_m = \rho \cos\left[\left(m+1-\kappa_m\frac{M}{3}\right)\frac{6\pi}{M}\right] \\ y_m = \rho \sin\left[\left(m+1-\kappa_m\frac{M}{3}\right)\frac{6\pi}{M}\right] \quad ; \quad m = 1, ..., M \\ z_m = \frac{L}{2}\left(\kappa_m - 1\right) \end{cases} \tag{23}$$

where $\kappa_m = \left\lfloor \frac{3(m-1)}{M} \right\rfloor$ ($\lfloor \cdot \rfloor$ being the *floor* operator) and $\rho = 3.2\ [\lambda]$. To avoid the so-called *inverse crime* [1], two different voxel-based discretization of $\mathcal{D}$ have been considered for the inverse ($N = 10 \times 10 \times 10$) and the forward ($N_{fwd} = 20 \times 20 \times 20$) problem, respectively. As for a quantitative evaluation of the reconstructions, the following error metric has been computed

$$\xi_R \triangleq \frac{1}{N_R} \sum_{n=1}^{N_R} \frac{\left|\widehat{\tau}^{(n)} - \tau^{(n)}\right|}{\left|\tau^{(n)}\right| + 1} \tag{24}$$

for the whole investigation domain ($R = tot \Rightarrow N_{tot} = N$), the scatterers support ($R = int \Rightarrow N_{int} = O$), and its complementary region (It is worthwhile to remark that the scatterer support is not an *a-priori* information exploited during the inversion, but it is rather employed in the post-processing phase only to compute the error metrics (24).) [$R = ext \Rightarrow N_{ext} = (N - O)$], $\tau^{(n)}$ and $\widehat{\tau}^{(n)}$ being the actual and the estimated contrast of the n-th ($n = 1, ..., N$) voxel in $\mathcal{D}$, respectively.

The first test case is concerned with the retrieval of a single ($K = 1$, K being the number of disconnected scattering regions lying in $\mathcal{D}$) off-centered scatterer composed by a single-voxel ($O = 1$) of side $\ell_x = \ell_y = \ell_z = 0.125\ [\lambda]$ with contrast $\tau = 1.0$ (Figure 2a). First, a sensitivity analysis for setting the optimal trade-off values of the control parameters of the *MT-BCS* solver [i.e., α and β in (19)] has been carried out. More specifically, the reconstruction error (24) has been computed by varying the values of the calibration coefficients within the ranges $\alpha \in \left[1, 10^2\right]$ and $\beta \in \left[10^{-5}, 10^{-2}\right]$ for different signal-to-noise ratios (*SNRs*). Figure 2b shows that the reconstruction "quality" is almost insensitive to the choice of α when $SNR > 5\ [dB]$ and it mainly depends on the noise level.

Differently, significant degradations occur when letting $\beta > 5 \times 10^{-4}$ whatever the *SNR* (Figure 2c). By computing the optimal *trade-off* value as $\varsigma^{(opt)} \triangleq \frac{\int_{SNR} \varsigma^{(opt)}\big|_{SNR} dSNR}{\int_{SNR} dSNR}$, $\varsigma^{(opt)}\big|_{SNR} \triangleq$ $\arg\left\{\min_\varsigma\left(\xi_{tot}\big|SNR\right)\right\}$ being the best value of $\varsigma = \{\alpha, \beta\}$ at a fixed *SNR* ($\cdot\big|$ standing for "evaluated at"), it turned out that $\alpha^{(opt)} = 10$ and $\beta^{(opt)} = 10^{-4}$. These thresholds have then been used throughout the whole numerical validation. A pictorial view of the 3D *MT-BCS* reconstructions when processing different noisy data is shown in Figure 3.

As it can be observed, the object support is always faithfully detected regardless of the amount of noise. There are only slight deviations from the actual contrast value (e.g., $\widehat{\tau}\big|_{SNR=50\,[dB]} = 0.81$—Figure 3a; $\widehat{\tau}\big|_{SNR=5\,[dB]} = 0.77$—Figure 3d) as quantitatively indicated by the corresponding values of the error index ξ_{tot} being equal to $\xi_{tot}\big|_{SNR=50\,[dB]} = 9.58 \times 10^{-5}$ (Figure 3a) and $\xi_{tot}\big|_{SNR=5\,[dB]} = 1.16 \times 10^{-4}$ (Figure 3d) at the highest and lowest *SNRs*, respectively.

In order to prove that such results are not due to a customization of the *MT-BCS* solver to the specific scenario at hand, a set of $W = 100$ inversions has been performed by randomly changing the position of the same target within the investigation domain. The results of such a statistical analysis are summarized in Table 1 where the minimum ($\xi_{tot}^{min} \triangleq \min_{w=1, ..., W}\{\xi_{tot,w}\}$), the maximum ($\xi_{tot}^{max} \triangleq \max_{w=1, ..., W}\{\xi_{tot,w}\}$), the average ($\xi_{tot}^{avg} \triangleq \frac{1}{W}\sum_{w=1}^{W}\xi_{tot,w}$), and the variance [$\xi_{tot}^{var} \triangleq \frac{1}{W}\sum_{w=1}^{W}\left(\xi_{tot,w} - \xi_{tot}^{avg}\right)^2$] values of the total error among the W experiments are reported. Whatever the *SNR*, the scatterer retrieval is very accurate since on average $\xi_{tot}^{avg} < 1.0 \times 10^{-4}$ (Table 1) and the variance of the reconstruction error, ξ_{tot}^{var}, is very small (i.e., $\xi_{tot}^{var} < 5.0 \times 10^{-11}$—Table 1).

Such a positive outcome holds true when dealing with stronger scatterers, as well. Thanks to the *CSI* formulation, which avoids any physical assumption/approximation in the application of the *CS* to the 3D scattering equations (Section 2), faithful data inversions are yielded also when increasing the actual contrast well-beyond the value of the first example. As a proof, the plots of ζ_{tot} (Figure 4a), ζ_{int} (Figure 4b), and ζ_{ext} (Figure 4c) versus τ and the noise level indicate that $\mathcal{D}$ can be carefully imaged when the scatterer is at least up to $\tau = 4.0$.

As expected, the reconstruction progressively degrades when higher and higher contrasts are at hand especially when processing highly blurred data. Figure 5 shows the actual and the retrieved volumetric distributions when $\tau = 4.0$ and $SNR \leq 10$ [dB]. As it can be inferred, some artifacts are present only in very harsh conditions (Figure 5c—$SNR = 5$ [dB]) when the external error (Figure 4c) grows from $\zeta_{ext}\rfloor_{SNR=5\,[dB]}^{\tau=1.0} = 0.0$ (Figure 3d) to $\zeta_{ext}\rfloor_{SNR=5\,[dB]}^{\tau=4.0} = 8.96 \times 10^{-6}$ (Figure 5c). Nevertheless, it is worth pointing out that these inaccuracies correspond to voxels with a very low contrast (i.e., $\widehat{\tau} < 3 \times 10^{-4}$—Figure 5c) that can be easily filtered out by a simple (i.e., the result is not so sensitive to the choice of the threshold value τ_{th}) thresholding (Figure 5d—$\tau_{th} = 10^{-3}$) and the scatterer is always correctly localized (Figure 5c as well as Figure 5d vs. Figure 5a).

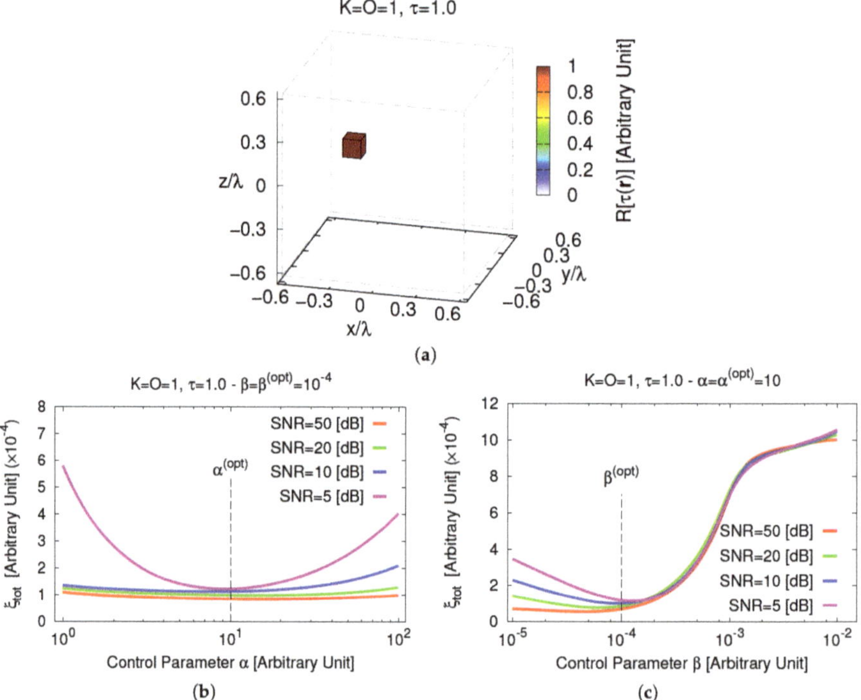

Figure 2. *Sensitivity Analysis* ($K = 1$, $O = 1$, $\tau = 1.0$, $SNR \in [5, 50]$ [dB])—Actual contrast function (**a**). Behavior of the total error, ξ_{tot}, versus the *MT-BCS* control parameters: (**b**) $\alpha \in [1, 10^2]$ ($\beta = \beta^{(opt)} = 10^{-4}$) and (**c**) $\beta \in [10^{-5}, 10^{-2}]$ ($\alpha = \alpha^{(opt)} = 10$).

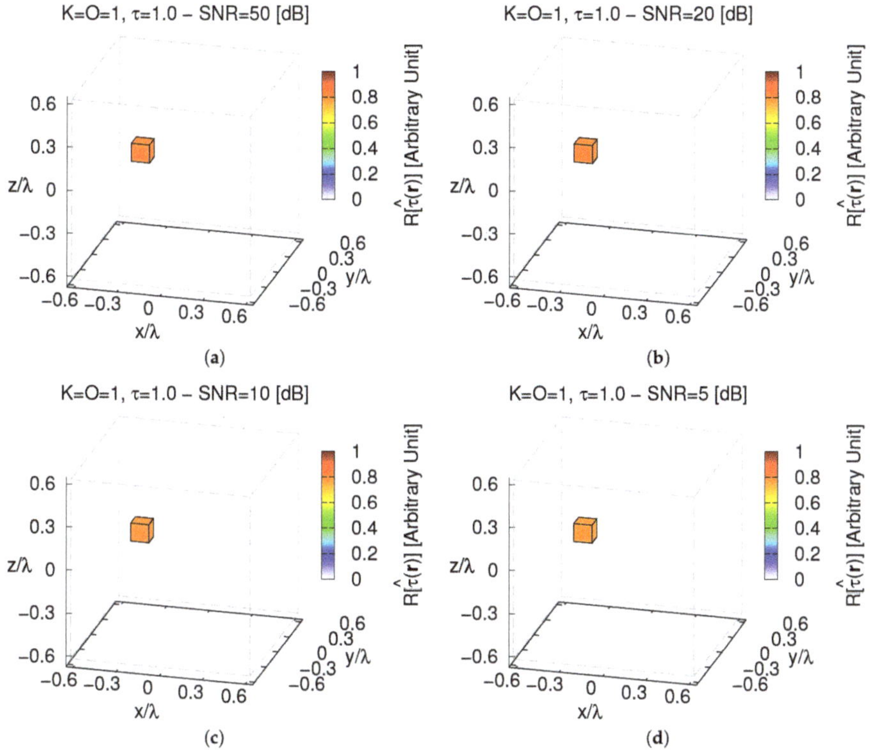

Figure 3. *Numerical Assessment* ($K = 1, O = 1, \tau = 1.0$) - *MT-BCS* reconstructions when processing the scattering data with (**a**) $SNR = 50$ [dB] ($\hat{\tau} = 0.81$); (**b**) $SNR = 20$ [dB] ($\hat{\tau} = 0.79$); (**c**) $SNR = 10$ [dB] ($\hat{\tau} = 0.78$), and (**d**) $SNR = 5$ [dB] ($\hat{\tau} = 0.77$).

Table 1. *Numerical Assessment* ($K = 1, O = 1, \tau = 1.0, W = 100, MT\text{-}BCS$)—Total error statistics.

SNR [dB]	ξ_{tot}^{min}	ξ_{tot}^{max}	ξ_{tot}^{avg}	ξ_{tot}^{var}
50	7.99×10^{-5}	9.69×10^{-5}	9.10×10^{-5}	1.11×10^{-11}
20	8.37×10^{-5}	9.75×10^{-5}	9.12×10^{-5}	1.21×10^{-11}
10	8.66×10^{-5}	9.86×10^{-5}	9.27×10^{-5}	1.87×10^{-11}
5	8.68×10^{-5}	1.21×10^{-4}	9.78×10^{-5}	4.66×10^{-11}

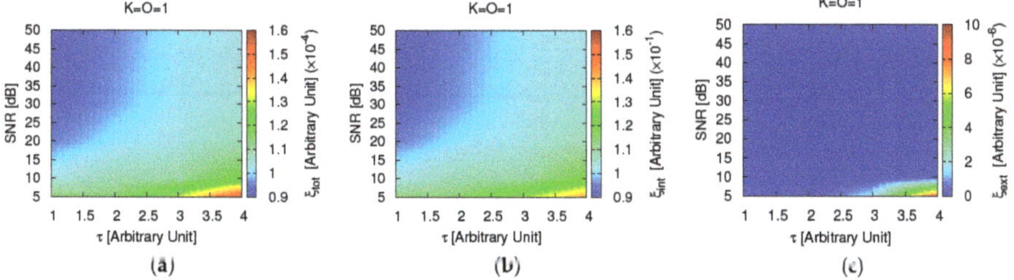

Figure 4. *Numerical Assessment* ($K = 1, O = 1, \tau \in [1.0, 4.0], SNR \in [5, 50]$ [dB])—Behavior of the (**a**) total (ξ_{tot}); (**b**) internal (ξ_{int}); and (**c**) external (ξ_{ext}) reconstruction errors when processing the scattering data with the *MT-BCS*.

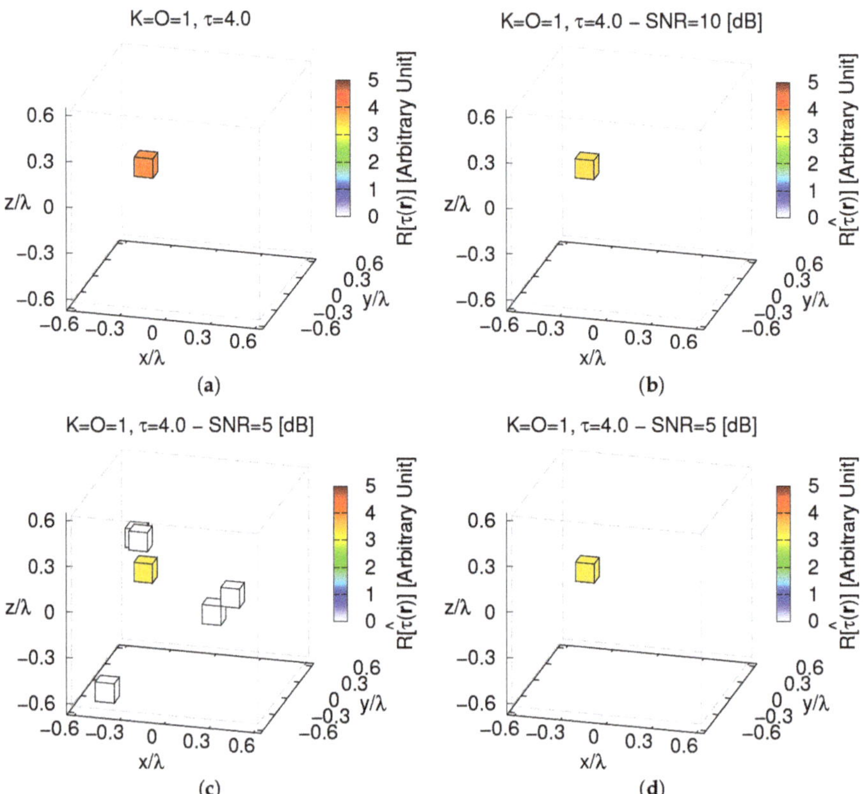

Figure 5. *Numerical Assessment* ($K = 1$, $O = 1$, $\tau = 4.0$)—(**a**) Actual contrast function and (**b,c**) *MT-BCS* reconstructions when processing the scattering data with (**b**) $SNR = 10$ [dB] ($\hat{\tau} = 3.40$), (**c**) $SNR = 5$ [dB] (unfiltered) ($\hat{\tau} = 3.28$), and (**d**) $SNR = 5$ [dB] (filtered $\tau_{th} = 10^{-3}$) ($\hat{\tau} = 3.28$).

The previous examples have dealt with real-valued target contrasts. However, the imaginary part of τ is not enforced to be zero during the inversion (see Section 3). In order to assess the reliability of the *MT-BCS* also when complex contrast values are at hand, the retrieval of a $K = 1$, $O = 1$ target with $\tau = 1.0 - 0.6j$ (Figure 6a,b) has been considered next. The analysis of the retrieved profiles (Figure 6) shows that the location, size, and contrast of the scatterer is correctly retrieved by the proposed imaging strategy both in low (e.g., $SNR = 50$ [dB]—Figure 6c,d) and in high noise conditions (e.g., $SNR = 5$ [dB]—Figure 6e,f), as shown by the corresponding error indexes (e.g., $\xi_{tot} \rfloor_{SNR=50\,[dB]} = 8.63 \times 10^{-5}$—Figure 6c,d; $\xi_{tot} \rfloor_{SNR=5\,[dB]} = 1.06 \times 10^{-4}$—Figure 6e,f).

Being assessed the effectiveness of the proposed approach in reconstructing the sparsest ($O = 1$) actual profile, let us now analyze its performance for scenarios exhibiting a lower *sparsity* order with respect to the selected voxel basis (7). Towards this end, a second single-voxel scatterer has been added to the configuration of the first test case (Figure 2a), but in a different position in $\mathcal{D}$ ($K = O = 2$, $\tau = 1.0$—Figure 7a).

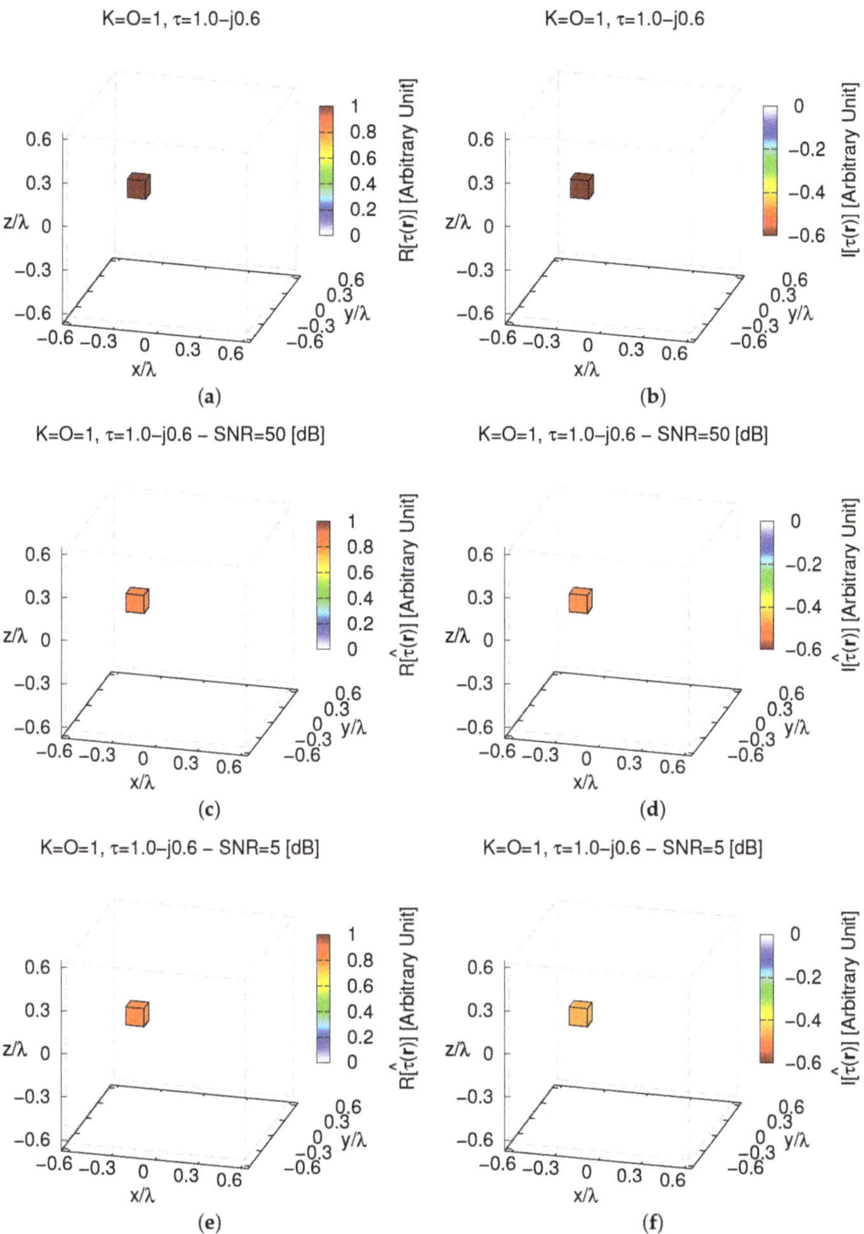

Figure 6. *Numerical Assessment* ($K = 1$, $O = 1$, $\tau = 1.0 - 0.6j$)—Real (**a,c,e**) and imaginary parts (**b,d,f**) of the (**a,b**) actual contrast function and of the (**b–f**) *MT-BCS* reconstructed profiles when processing the scattering data with (**c,d**) $SNR = 50$ [dB] ($\hat{\tau} = 0.84 - 0.49j$), (**e,f**) $SNR = 5$ [dB] ($\hat{\tau} = 0.81 - 0.45j$).

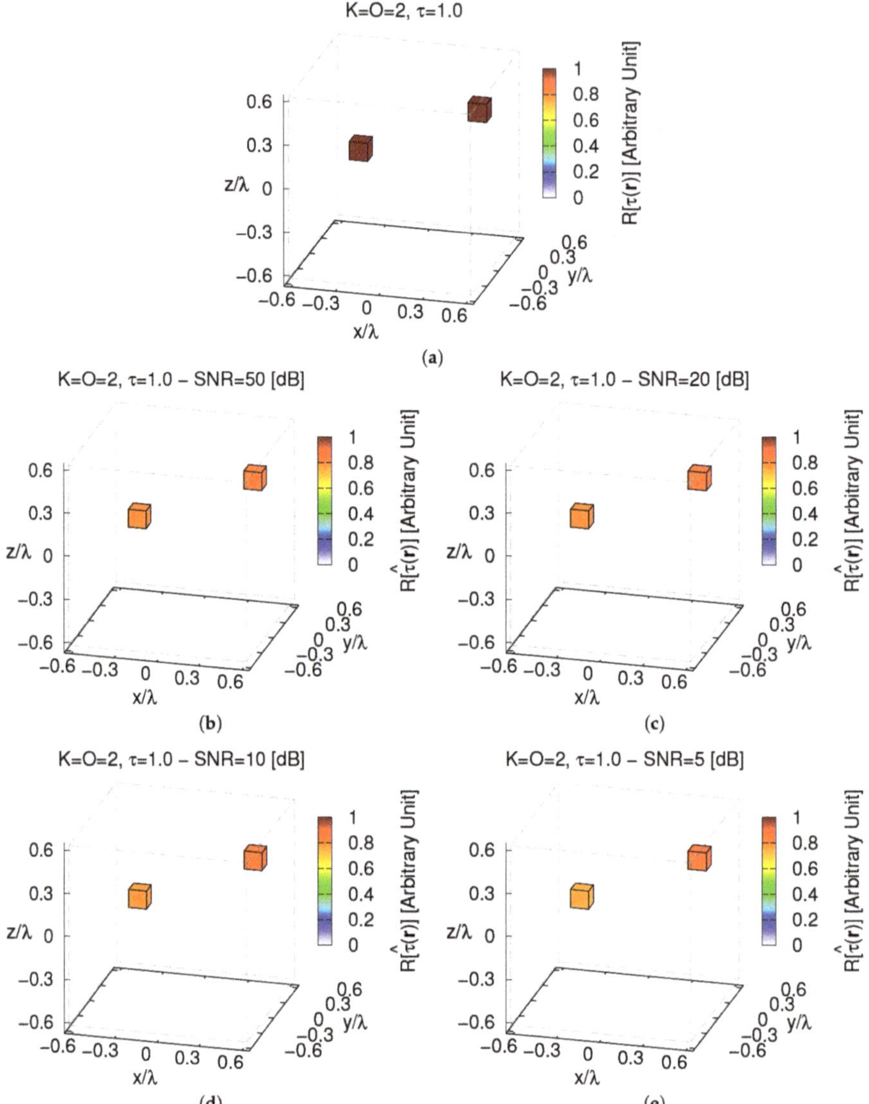

Figure 7. *Numerical Assessment (K = 2, O = 2, τ = 1.0)*—(**a**) Actual contrast function and *MT-BCS* reconstructions when processing the scattering data with (**b**) $SNR = 50$ [dB] ($\hat{\tau}_{\max} = 0.82$), (**c**) $SNR = 20$ [dB] ($\hat{\tau}_{\max} = 0.84$), (**d**) $SNR = 10$ [dB] ($\hat{\tau}_{\max} = 0.84$), and (**e**) $SNR = 5$ [dB] ($\hat{\tau}_{\max} = 0.83$).

As it can be inferred from the plots in Figure 7b–e, the *MT-BCS* is always able to correctly retrieve both scatterers even from very low-noise data (e.g., $\zeta_{tot}\rfloor_{SNR=5\,[\text{dB}]} = 9.80 \times 10^{-4}$—Figure 7e). However, the same statistical analysis of the single-scatterer/single-voxel case (i.e., randomly changing the locations of the scatterers) here results in wider variations of the integral error (i.e., $\zeta_{tot}^{var} \geq 4.5 \times 10^{-7}$ (Table 2) vs. $\zeta_{tot}^{var} \geq 1.11 \times 10^{-11}$ (Table 1)). Therefore, to better understand how the inversion accuracy is affected by the positions of the scatterers within the investigation domain, a further set of experiments has been performed by deterministically changing the relative locations of the $K = 2$ scatterers, d being the distance of each one of them from the origin.

Table 2. *Numerical Assessment* ($K = 2, O = 2, \tau = 1.0, W = 100, MT\text{-}BCS$)—Total error statistics.

SNR [dB]	ζ_{tot}^{min}	ζ_{tot}^{max}	ζ_{tot}^{avg}	ζ_{tot}^{var}
50	1.49×10^{-4}	2.57×10^{-3}	4.60×10^{-4}	4.50×10^{-7}
20	1.56×10^{-4}	2.59×10^{-3}	4.63×10^{-4}	4.96×10^{-7}
10	1.61×10^{-4}	2.62×10^{-3}	4.72×10^{-4}	5.20×10^{-7}
5	1.83×10^{-4}	2.63×10^{-3}	5.30×10^{-4}	5.27×10^{-7}

Some representative test cases are reported in Figure 8: $d = d_{min} = \ell \frac{\sqrt{3}}{2} = 0.11$ [λ] (Figure 8a,b), $d = 0.54$ [λ] (Figure 8c,d), and $d = d_{max} = (L - \ell) \frac{\sqrt{3}}{2} = 0.97$ [λ] (Figure 8e,f). As expected, the reconstruction worsens when the two objects get closer (Figure 9) until some artifacts appear when $d = d_{min}$ [$\frac{\zeta_{tot}\rfloor_{d=0.11\,[\lambda]}^{SNR=10\,[dB]}}{\zeta_{tot}\rfloor_{d=0.97\,[\lambda]}^{SNR=10\,[dB]}} \simeq 2.41$—Figure 8b vs. Figure 8f].

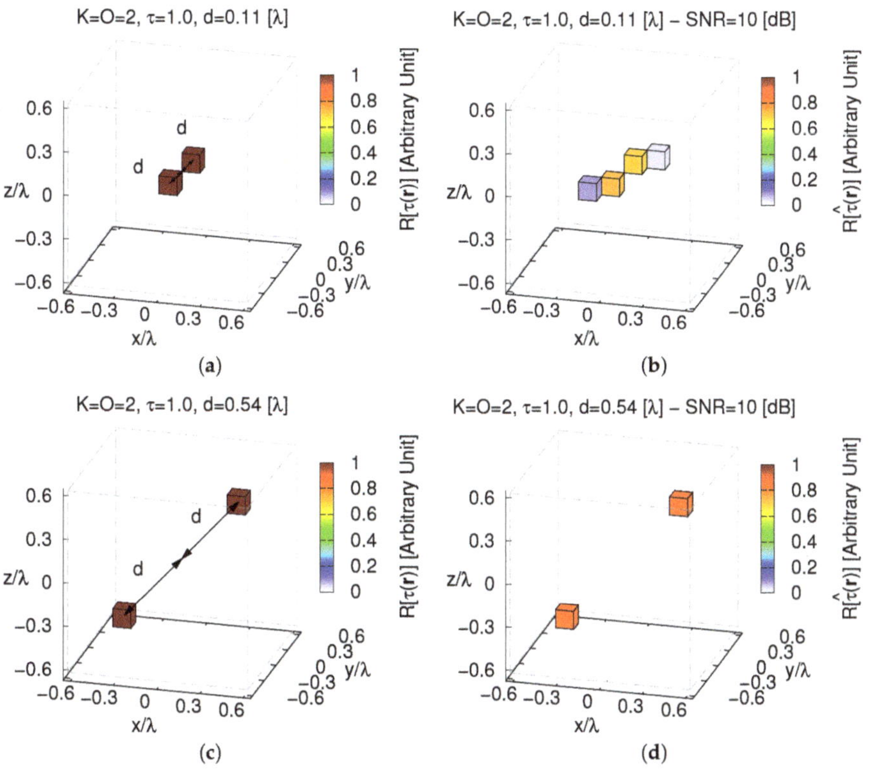

Figure 8. *Cont.*

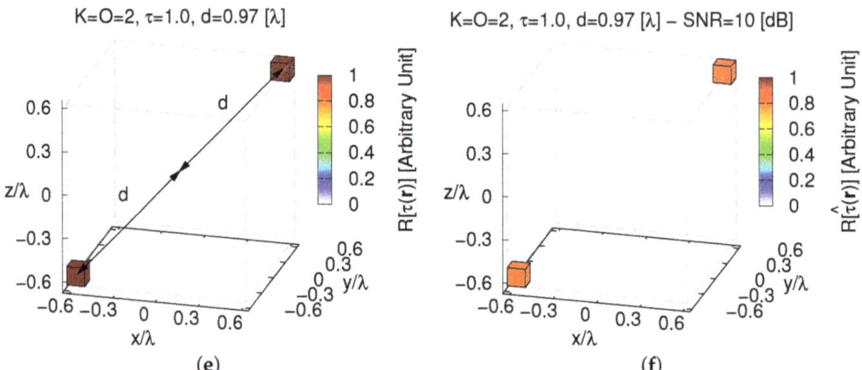

Figure 8. *Numerical Assessment* ($K = 2$, $O = 2$, $\tau = 1.0$, $SNR = 10$ [dB])—(**a,c,e**) Actual contrast function and (**b,d,f**) *MT-BCS* reconstructions when the distance of the scatterers from the origin is equal to (**a,b**) $d = d_{\min} = 0.11$ [λ] ($\hat{\tau}_{\max} = 0.73$), (**c,d**) $d = 0.54$ [λ] ($\hat{\tau}_{\max} = 0.83$), and (**e,f**) $d = d_{\max} = 0.97$ [λ] ($\hat{\tau}_{\max} = 0.84$).

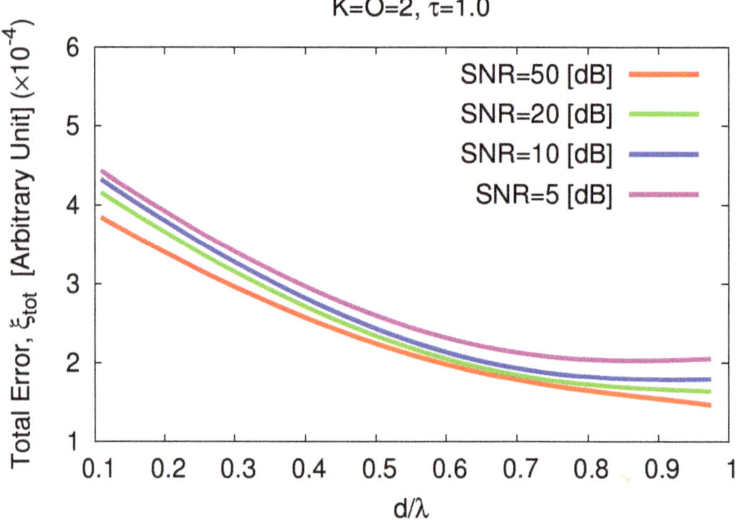

Figure 9. *Numerical Assessment* ($K = 2$, $O = 2$, $\tau = 1.0$, $SNR \in [5, 50]$ [dB])—Behavior of the total error, ξ_{tot}, versus the distance of the scatterers from the origin, d.

Further lowering the *sparseness* of the actual profile causes a decrement of the *CS* performance as proven by the results of the third test case when $O = 4$ and the $K = 4$ single-voxel scatterers are randomly placed in $\mathcal{D}$ (Figure 10a).

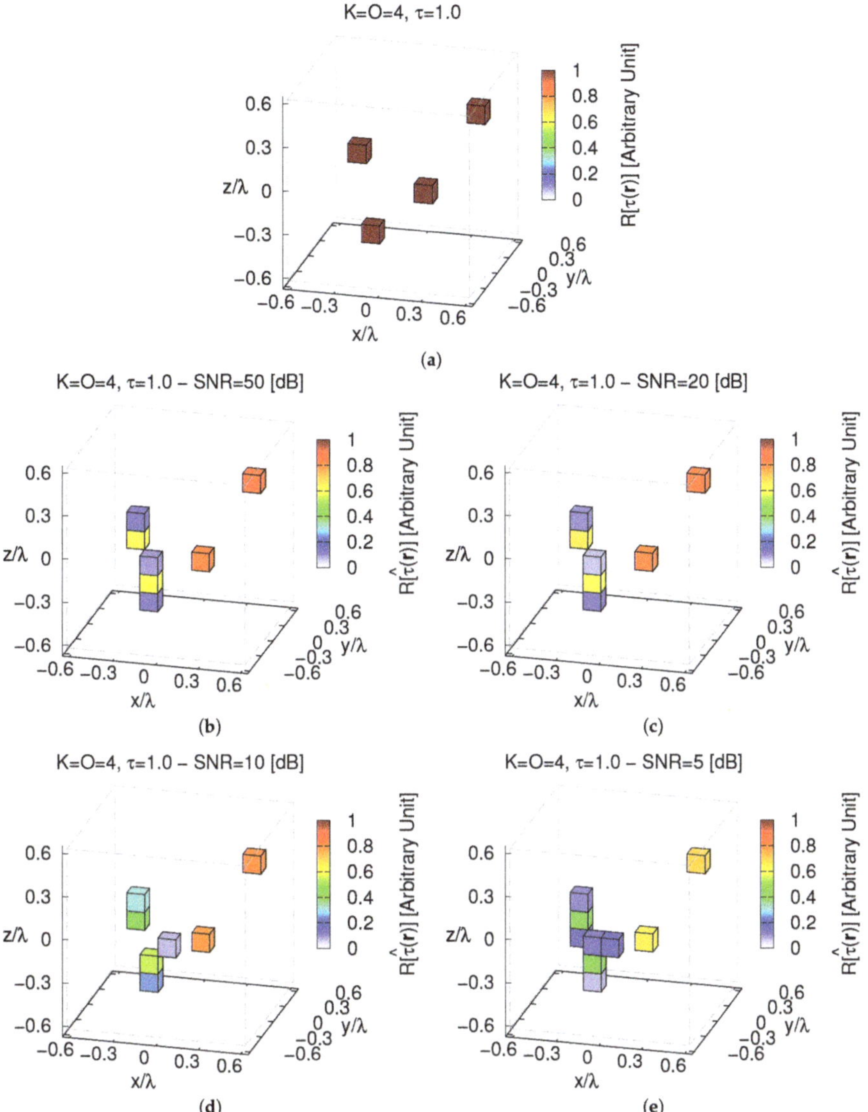

Figure 10. *Numerical Assessment* ($K = 4$, $O = 4$, $\tau = 1.0$)—(**a**) Actual contrast function and *MT-BCS* reconstructions when processing the scattering data characterized by (**b**) $SNR = 50$ [dB] ($\hat{\tau}_{max} = 0.84$), (**c**) $SNR = 20$ [dB] ($\hat{\tau}_{max} = 0.84$), (**d**) $SNR = 10$ [dB] ($\hat{\tau}_{max} = 0.80$), and (**e**) $SNR = 5$ [dB] ($\hat{\tau}_{max} = 0.70$).

Unavoidably, the *MT-BCS* gets worse and undesired artifacts occur also when processing low-noisy data (e.g., $\zeta_{tot}\rvert_{SNR=50\,[dB]} = 2.37 \times 10^{-3}$—Figure 10b), but still without preventing the possibility to correctly identify $K = 4$ separate objects (Figure 10b–e). In addition, in this case, changing the distance d of the objects with respect to the origin influences the quality of the retrieval as pictorially shown in Figure 11 ($SNR = 10$ [dB]).

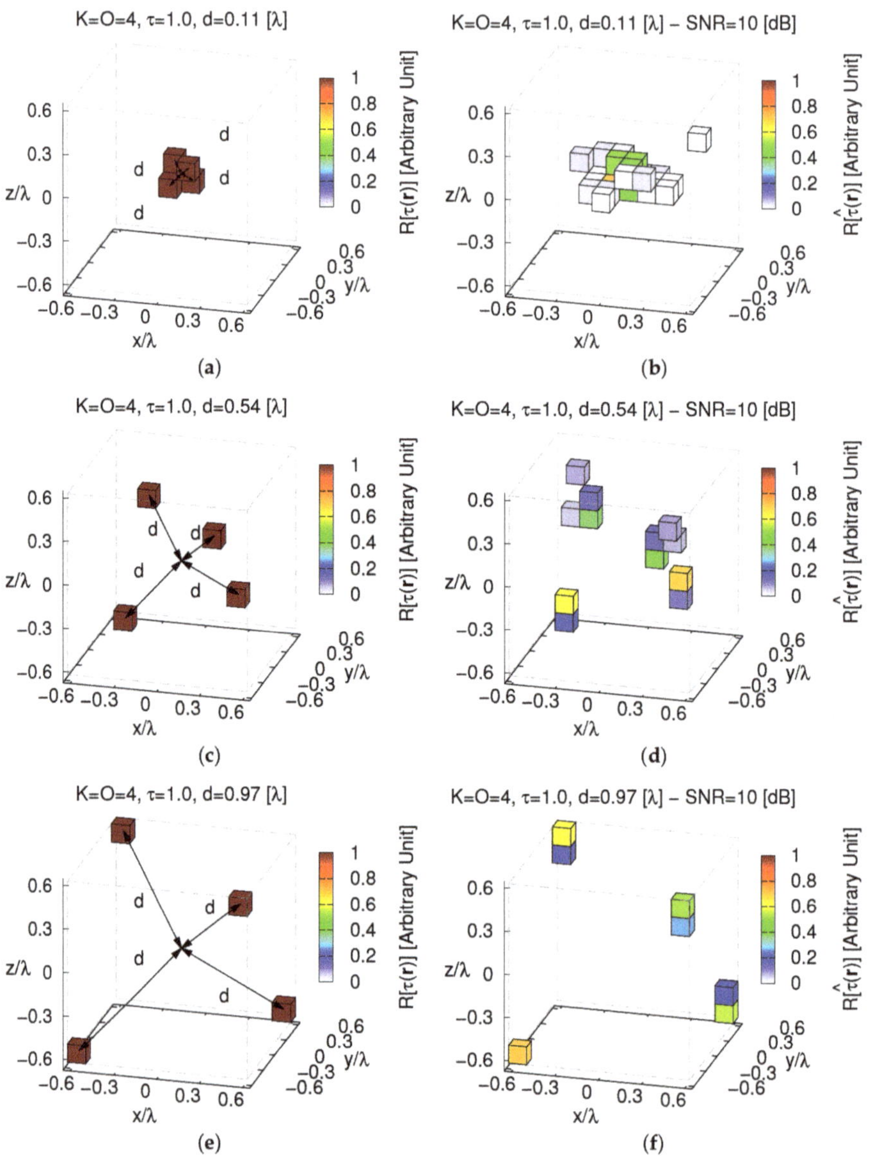

Figure 11. *Numerical Assessment* ($K = 4$, $O = 4$, $\tau = 1.0$, $SNR = 10$ [dB])—(**a,c,e**) *Actual contrast function and* (**b,d,f**) *MT-BCS reconstructions when the objects distance from the origin is* (**a,b**) $d = d_{\min} = 0.11$ [λ] ($\hat{\tau}_{\max} = 0.73$), (**c,d**) $d = 0.54$ [λ] ($\hat{\tau}_{\max} = 0.70$), *and* (**e,f**) $d = d_{\max} = 0.97$ [λ] ($\hat{\tau}_{\max} = 0.71$).

Indeed, it turns out that more accurate images of $\mathcal{D}$ are yielded when the scatterers are far (i.e., $d \uparrow \Rightarrow \xi_{tot} \downarrow$) as confirmed by the plot of ξ_{tot} vs. d (Figure 12a). However, it cannot be neglected that a simple filtering (Figure 12b—$\tau_{th} = 10^{-3}$) allows one to clearly resolve the scatterer support even in the most critical case (Figure 11a).

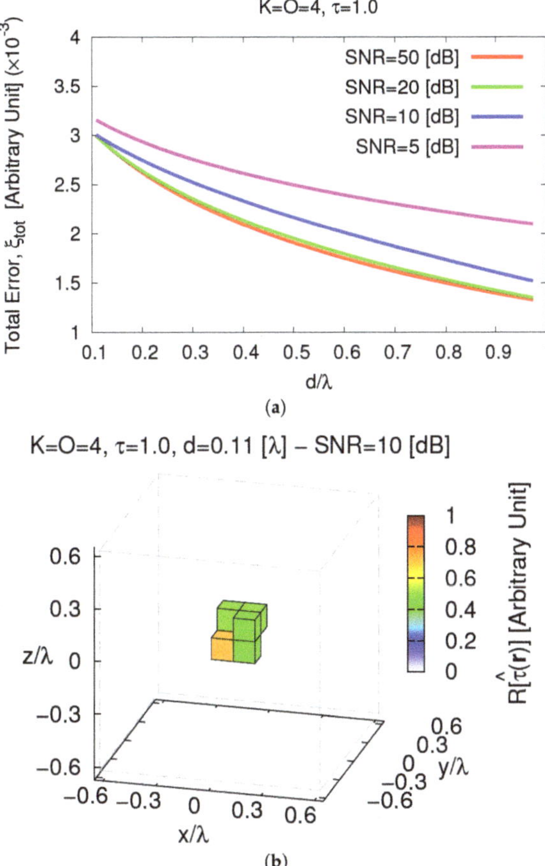

Figure 12. *Numerical Assessment* ($K = 4, O = 4, \tau = 1.0, SNR \in [5, 50]$ *[dB]*)—(**a**) Behavior of the total error, ξ_{tot}, as a function of the objects distance from the origin, d and (**b**) filtered ($\tau_{th} = 10^{-3}$) *MT-BCS* reconstruction when the objects distance from the origin is $d = d_{\min} = 0.11$ [λ] ($\hat{\tau}_{\max} = 0.73$).

The next numerical test is devoted to validate the *MT-BCS* in a more complex *3D* imaging scenario concerned with non-uniformly shaped scatterers. More specifically, $K = 3$ objects sized $\ell_x^1 = \ell_y^1 = \ell_z^1 = 0.125$ [λ], $\left[\ell_x^2, \ell_y^2, \ell_z^2\right] = [0.125, 0.25, 0.125]$ [λ], and $\left[\ell_x^3, \ell_y^3, \ell_z^3\right] = [0.25, 0.25, 0.125]$ [λ] ($O = 7$) with contrast $\tau = 1.0$ have been imaged (Figure 13a).

Despite the increased complexity and the reduced intrinsic-sparsity order of the scattering configuration, faithful reconstructions of the 3*D* contrast distribution are yielded (Figure 13b–e) even though there is an over-estimation of the scatterers supports when highly-blurred data are at hand (e.g., Figure 13b vs. Figure 13e) and, consequently, an increase of the reconstruction error with the noise level (e.g., $\frac{\xi_{tot}\rfloor^{SNR=5\,[dB]}}{\xi_{tot}\rfloor^{SNR=50\,[dB]}} \approx 2.04$). Moreover, the previous considerations regarding the fidelity of the proposed strategy with respect to the target contrast are confirmed also when dealing with non-uniformly shaped scatterers (Figure 14).

Finally, it is interesting to underline the advantage of using the *MT* extension of the *DCS*-based approach for solving the 3*D-CSI* inversion problem instead of its "naive" single-task implementation (*ST-BCS* [23]) that does not impose any physical correlation in solving (14). Towards this end, let us consider the benchmark $O = 6$ voxel arrangement in Figure 15a with two ($K = 2$) "L-shaped" homogeneous ($\tau = 2.0$) objects.

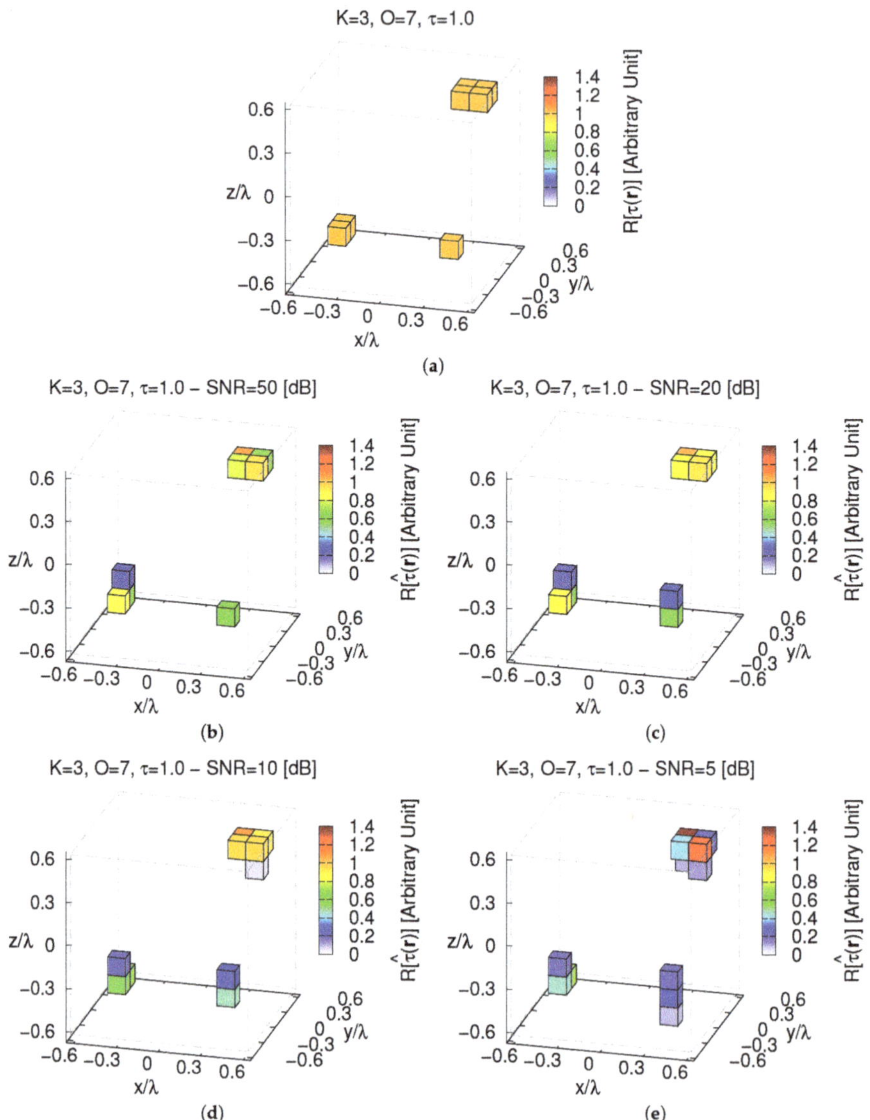

Figure 13. *Numerical Assessment* ($K = 3$, $O = 7$, $\tau = 1.0$)—(**a**) Actual contrast function and *MT-BCS* reconstructions when (**b**) $SNR = 50$ [dB] ($\hat{\tau}_{\max} = 1.08$); (**c**) $SNR = 20$ [dB] ($\hat{\tau}_{\max} = 1.05$); (**d**) $SNR = 10$ [dB] ($\hat{\tau}_{\max} = 1.09$); and (**e**) $SNR = 5$ [dB] ($\hat{\tau}_{\max} = 1.37$).

For illustrative purposes, the reconstructions with the *MT-BCS* (Figure 15b,c) and the *ST-BCS* (Figure 15d,e) when processing two different sets of noisy data [$SNR = 20$ [dB]—Figure 15b,d; $SNR = 10$ [dB]—Figure 15c,e] are shown. As it can be visually inferred, the *MT* strategy turns out to be more effective than the *ST* one in both locating and shaping the non-connected scattering regions as well as in estimating the actual contrast value. Such an inference is quantitatively confirmed by the error indexes since $\frac{\xi_{tot}\rfloor_{ST-BCS}^{SNR=20\,[dB]}}{\xi_{tot}\rfloor_{MT-BCS}^{SNR=20\,[dB]}} \approx 33.36$ (Figure 15b vs. Figure 15d) and $\frac{\xi_{tot}\rfloor_{ST-BCS}^{SNR=10\,[dB]}}{\xi_{tot}\rfloor_{MT-BCS}^{SNR=10\,[dB]}} \approx 7.04$

(Figure 15c vs. Figure 15e) (Table 3). Similar outcomes can also be drawn when changing the contrast of the scatterers as indicated by the behaviour of the reconstruction error ξ_{tot} versus τ in Figure 16.

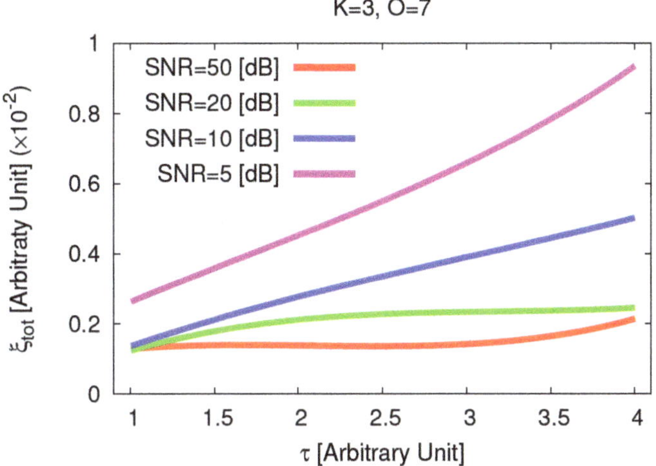

Figure 14. *Numerical Assessment ($K = 3$, $O = 7$, $SNR \in [5, 50]$ [dB])*—Behavior of the total reconstruction error (ξ_{tot}) when processing the scattering data with the *MT-BCS*.

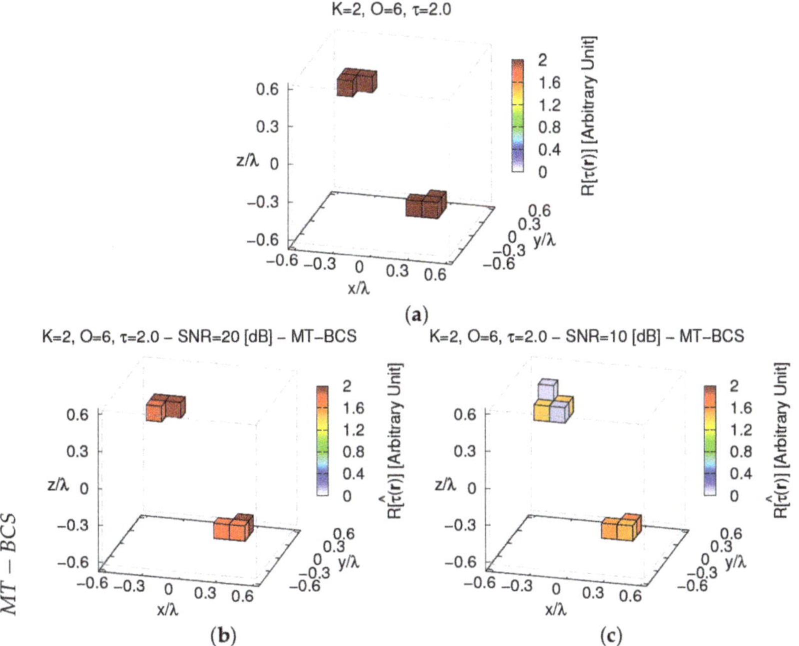

Figure 15. *Cont.*

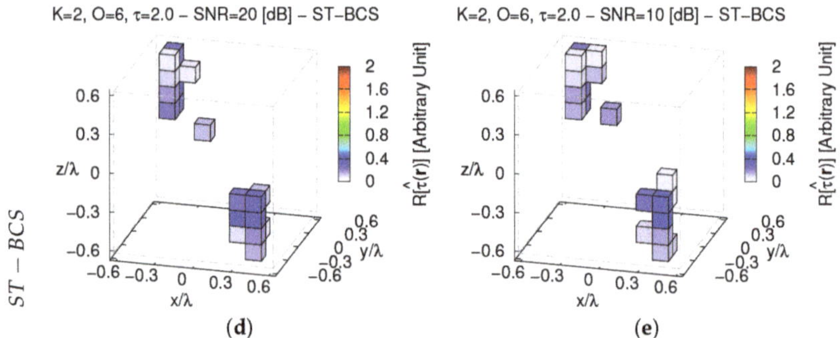

Figure 15. *Comparative Assessment* ($K = 2, O = 6, \tau = 2.0$)—(**a**) Actual contrast function and retrieved solutions by the (**b,c**) *MT-BCS* and (**d,e**) *ST-BCS* when processing noisy data at (**b,d**) $SNR = 20$ [dB] ($\hat{\tau}_{max}^{MT-BCS} = 1.96$, $\hat{\tau}_{max}^{ST-BCS} = 0.33$) and (**c,e**) $SNR = 10$ [dB] ($\hat{\tau}_{max}^{MT-BCS} = 1.68$, $\hat{\tau}_{max}^{ST-BCS} = 0.36$).

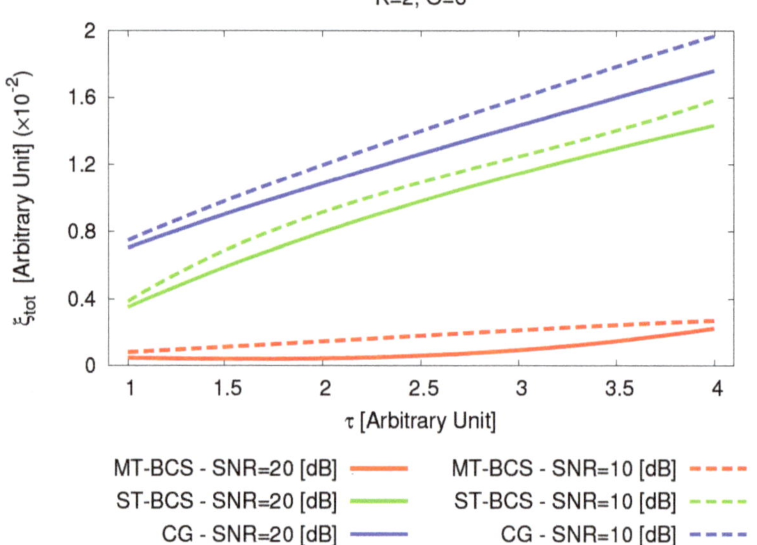

Figure 16. *Comparative Assessment* ($K = 2, O = 6, \tau \in [1.0, 4.0]$, $SNR \in [10, 20]$ [dB])—Behavior of the total error, ξ_{tot}, as a function of the object contrast, τ, when processing the scattering data with the *MT-BCS*, the *ST-BCS*, and the *CG* methods.

To conclude the numerical assessment of the reconstruction capabilities of the *MT-BCS*, it has been compared with a competitive non-*CS* state-of-the-art approach. Towards this end, a deterministic *CG*-based inversion tool—still based on a *CSI* formulation of the scattering problem—has been applied to the same scenario in Figure 15a. The retrieved images of the dielectric profile of the investigation domain are shown in Figure 17a ($SNR = 20$ [dB]) and Figure 17b ($SNR = 10$ [dB]).

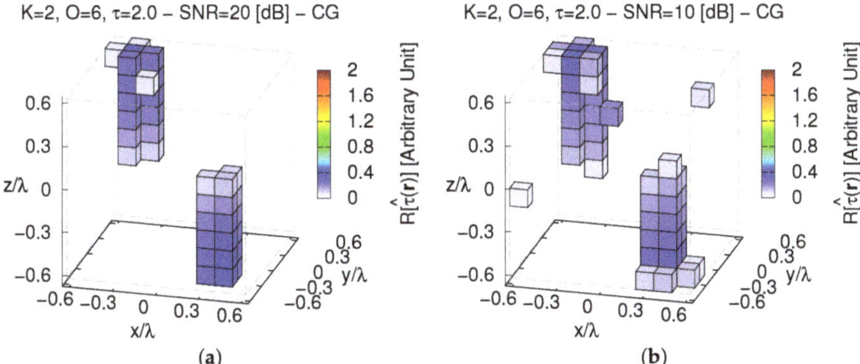

Figure 17. *Comparative Assessment* ($K = 2, O = 6, \tau = 2.0$)—*CG* reconstructions when processing noisy data characterized by (**a**) $SNR = 20$ [dB] ($\hat{\tau}^{CG}_{max} = 0.33$) and (**b**) $SNR = 10$ [dB] ($\hat{\tau}^{CG}_{max} = 0.36$).

Without imposing *sparseness* priors, only the presence of $K = 2$ scatterers lying in $\mathcal{D}$ can be deduced, but their contrasts are strongly under-estimated and their supports/shapes are unreliably predicted. Comparatively, the *MT-BCS* enables a reduction of the reconstruction error of about $\frac{\xi_{tot}\rfloor^{SNR=20\,[dB]}_{CG}}{\xi_{tot}\rfloor^{SNR=20\,[dB]}_{MT-BCS}} \approx 43.75$ (Figure 17a vs. Figure 15b—Table 3) and $\frac{\xi_{tot}\rfloor^{SNR=10\,[dB]}_{CG}}{\xi_{tot}\rfloor^{SNR=10\,[dB]}_{MT-BCS}} \approx 8.92$ (Figure 17b vs. Figure 15c—Table 3) times. Such conclusions are not limited to the contrast $\tau = 1.0$, but they also arise for stronger scatterers as detailed by the plot of ξ_{tot} versus τ in Figure 16.

As for the computational efficiency, the *MT*-constrained exploitation of a sparseness promoting inversion technique allows a non-negligible reduction of the computational time, Δt, when processing 3D scattering data (For the sake of fairness, all the computational times refer to non-optimized Matlab codes executed on a single-core laptop running at 2.20 GHz). Indeed, the *MT-BCS* is not only faster than the *ST-BCS*, thanks to the joint processing of the data (i.e., $\frac{\Delta t\rfloor_{MT-BCS}}{\Delta t\rfloor_{ST-BCS}} \approx 0.52$—Table 3), but it also overcomes the *CG* speed (i.e., $\frac{\Delta t\rfloor_{MT-BCS}}{\Delta t\rfloor_{CG}} \approx 0.032$—Table 3).

Table 3. *Comparative Assessment* ($K = 2, O = 6, \tau = 2.0, SNR \in [10, 20]$ [dB])—Inversion performance indexes.

Method	$SNR = 20$ [dB]			$SNR = 10$ [dB]			Δt
	ξ_{tot}	ξ_{int}	ξ_{ext}	ξ_{tot}	ξ_{int}	ξ_{ext}	[s]
$MT - BCS$	2.56×10^{-4}	4.07×10^{-2}	0.00	1.39×10^{-3}	1.53×10^{-1}	3.94×10^{-4}	15.12
$ST - BCS$	8.54×10^{-3}	6.28×10^{-1}	2.95×10^{-3}	9.79×10^{-3}	8.16×10^{-1}	3.06×10^{-3}	29.08
CG	1.12×10^{-2}	5.78×10^{-1}	7.79×10^{-3}	1.24×10^{-2}	5.91×10^{-1}	8.92×10^{-3}	4.66×10^{2}

5. Conclusions

An innovative approach to efficiently solve the full-vectorial *3D-IS* problem has been presented. The retrieval of the volumetric contrast distribution of sparse non-weak scatterers has been tackled as a probabilistic *CSI*-based problem, which has been efficiently solved through a customized *MT-BCS* approach. The numerical analysis has pointed out the following key features of the proposed technique:

- Reliable *3D* reconstructions of the *EM* properties of the imaged domain are yielded processing scattering data also blurred with a non-negligible amount of additive noise;
- The inversion accuracy of the proposed *CS*-based approach depends on the degree of *sparseness* of the actual scenario with respect to the expansion basis at hand. However, it can be

fruitfully and profitably applied when other/different (non-voxel) representations of the contrast source/contrast function are chosen [46];

- The *MT* implementation of the *BCS*-based inversion remarkably overcomes its single-task (*ST-BCS*) counterpart thanks to the profitable exploitation of the existing correlations between the *V* views and the scattered field components;
- The *MT-BCS* positively compares with other state-of-the-art approaches, also deterministic and non-*CS*, in terms of both reconstruction accuracy and computational efficiency.

Moreover, the methodological advancements of this work with respect to the state-of-the-art on the topic [47,48] include (*i*) the generalization of the *MT-BCS* strategy to handle *3D-IS* problems, differently from previous customizations of such an inversion paradigm which deal only with two-dimensional formulations [48], (*ii*) the derivation of a *BCS*-based imaging approach able to retrieve *3D* target contrast information, unlike state-of-the-art *Bayesian CS* contributions only dealing with the reconstruction of *equivalent sources* [47], and (*iii*) the analysis and validation of suitable operative guidelines for the optimal setting of the key calibration parameters of the introduced methodology. Future works will be aimed at extending the capabilities of the proposed approach to effectively deal with non voxel-sparse targets as well as with other applicative scenarios of great interest (e.g., subsurface imaging) including the processing of multi-frequency data [55].

Author Contributions: All authors contributed equally to this work.

Funding: This work has been partially supported by the Italian Ministry of Foreign Affairs and International Cooperation, Directorate General for Cultural and Economic Promotion and Innovation within the SNATCH Project (2017–2019) and by the Italian Ministry of Education, University, and Research within the Program "Smart cities and communities and Social Innovation" (CUP: E44G14000060008) for the Project "WATERTECH—Smart Community per lo Sviluppo e l'Applicazione di Tecnologie di Monitoraggio Innovative per le Reti di Distribuzione Idrica negli usi idropotabili ed agricoli" (Grant no. SCN_00489).

Conflicts of Interest: The authors declare no conflict of interest.

References

1. Chen, X. *Computational Methods for Electromagnetic Inverse Scattering*; Wiley-IEEE: Singapore, 2018.
2. Zoughi, R. *Microwave Nondestructive Testing and Evaluation*; Kluwer: Amsterdam, The Netherlands, 2000.
3. Ghasr, M.T.; Horst, M.J.; Dvorsky, M.R.; Zoughi, R. Wideband microwave camera for real-time 3-D imaging. *IEEE Trans. Antennas Propag.* **2017**, *65*, 258–268. [CrossRef]
4. Fallahpour, M.; Zoughi, R. Fast 3-D qualitative method for through-wall imaging and structural health monitoring. *IEEE Geosci. Remote Sens. Lett.* **2015**, *12*, 2463–2467. [CrossRef]
5. Benjamin, R.; Craddock, I.J.; Hilton, G.S.; Litobarski, S.; McCutcheon, E.; Nilavalan, R.; Crisp, G.N. Microwave detection of buried mines using non-contact, synthetic near-field focusing. *IEE P-Radar Son. Nav.* **2001**, *148*, 233–240. [CrossRef]
6. Sheen, D.M.; McMakin, D.L.; Hall, T.E. Three-dimensional millimeter-wave imaging for concealed weapon detection. *IEEE Trans. Microwave Theory Technol.* **2001**, *49*, 1581–1592. [CrossRef]
7. Di Donato, L.; Crocco, L. Model-based quantitative cross-borehole GPR imaging via virtual experiments. *IEEE Trans. Geosci. Remote Sens.* **2015**, *53*, 4178–4185. [CrossRef]
8. Catapano, I.; Crocco, L.; Persico, R.; Pieraccini, M.; Soldovieri, F. Linear and nonlinear microwave tomography approaches for subsurface prospecting: Validation on real data. *IEEE Antennas Wirel. Propag. Lett.* **2006**, *5*, 49–53. [CrossRef]
9. Bucci, O.M.; Crocco, L.; Isernia, T.; Pascazio, V. Subsurface inverse scattering problems: Quantifying, qualifying, and achieving the available information. *IEEE Trans. Geosci. Remote Sens.* **2001**, *39*, 2527–2538. [CrossRef]
10. Bevacqua, M.; Crocco, L.; Di Donato, L.; Isernia, T.; Palmeri, R. Exploiting sparsity and field conditioning in subsurface microwave imaging of nonweak buried targets. *Radio Sci.* **2016**, *51*, 301–310. [CrossRef]
11. Amineh, R.K.; Khalatpour, A.; Nikolova, N.K. Three-dimensional microwave holographic imaging using co- and cross-polarized data. *IEEE Trans. Antennas Propag.* **2012**, *60*, 3526–3531. [CrossRef]

12. Semenov, S.Y.; Bulyshev, A.E.; Abubakar, A.; Posukh, V.G.; Sizov, Y.E.; Souvorov, A.E.; van den Ber, P.M.; Williams, T.C. Microwave-tomographic imaging of the high dielectric-contrast objects using different image-reconstruction approaches. *IEEE Trans. Microw. Theory Technol.* **2005**, *53*, 2284–2294. [CrossRef]

13. Semenov, S.Y.; Bulyshev, A.E.; Souvorov, A.E.; Nazarov, A.G.; Sizov, Y.E.; Svenson, R.H.; Posukh, V.G.; Pavlovsky, A.; Repin, P.N.; Tatsis, G.P. Three-dimensional microwave tomography: Experimental imaging of phantoms and biological objects. *IEEE Trans. Microwave Theory Technol.* **2000**, *48*, 1071–1074. [CrossRef]

14. Zhang, Z.Q.; Liu, Q.H. Three-dimensional nonlinear image reconstruction for microwave biomedical imaging. *IEEE Trans. Biomed. Eng.* **2004**, *51*, 544–548. [CrossRef] [PubMed]

15. Bulyshev, A.E.; Semenov, S.Y.; Souvorov, A.E.; Svenson, R.H.; Nazarov, A.G.; Sizov, Y.E.; Tatsis, G.P. Computational modeling of three-dimensional microwave tomography of breast cancer. *IEEE Trans. Biomed. Eng.* **2001**, *48*, 1053–1056. [CrossRef]

16. Colgan, T.J.; Hagness, S.C.; Van Veen, B.D. A 3-D level set method for microwave breast imaging. *IEEE Trans. Biomed. Eng.* **2015**, *62*, 2526–2534. [CrossRef]

17. Winters, D.W.; Shea, J.D.; Kosmas, P.; Van Veen, B.D.; Hagness, S.C. Three-dimensional microwave breast imaging: dispersive dielectric properties estimation using patient-specific basis functions. *IEEE Trans. Med. Imaging* **2009**, *28*, 969–981. [CrossRef] [PubMed]

18. Grzegorczyk, T.M.; Meaney, P.M.; Kaufman, P.A.; di Florio-Alexander, R.M.; Paulsen, K.D. Fast 3-D tomographic microwave imaging for breast cancer detection. *IEEE Trans. Med. Imaging* **2012**, *31*, 1584–1592. [CrossRef] [PubMed]

19. Johnson, J.E.; Takenaka, T.; Ping, K.A.H.; Honda, S.; Tanaka, T. Advances in the 3-D forward-backward time-stepping (FBTS) inverse scattering technique for breast cancer detection. *IEEE Trans. Biomed. Eng.* **2009**, *56*, 2232–2243. [CrossRef] [PubMed]

20. Bevacqua, M.T.; Scapaticci, R. A compressive sensing approach for 3D breast cancer microwave imaging with magnetic nanoparticles as contrast agent. *IEEE Trans. Med. Imaging* **2016**, *35*, 665–673. [CrossRef]

21. Bucci, O.M.; Bellizzi, G.; Catapano, I.; Crocco, L.; Scapaticci, R. MNP enhanced microwave breast cancer imaging: Measurement constraints and achievable performances. *IEEE Antennas Wirel. Propag. Lett.* **2012**, *11*, 1630–1633. [CrossRef]

22. Angiulli, G.; Carlo, D.D.; Isernia, T. Matching fluid influence on field scattered from breast tumour: Analysis using 3D realistic numerical phantoms. *Electron. Lett.* **2012**, *48*, 13–14. [CrossRef]

23. Oliveri, G.; Rocca, P.; Massa, A. A Bayesian compressive sampling-based inversion for imaging sparse scatterers. *IEEE Trans. Geosci. Remote Sens.* **2011**, *49*, 3993–4006. [CrossRef]

24. Palmeri, R.; Bevacqua, M.T.; Crocco, L.; Isernia, T.; Di Donato, L. Microwave imaging via distorted iterated virtual experiments. *IEEE Trans. Antennas Propag.* **2017**, *65*, 829–838. [CrossRef]

25. Di Donato, L.; Palmeri, R.; Sorbello, G.; Isernia, T.; Crocco, L. A new linear distorted-wave inversion method for microwave imaging via virtual experiments. *IEEE Trans. Microwave Theory Technol.* **2016**, *64*, 2478–2488. [CrossRef]

26. Di Donato, L.; Bevacqua, M.T.; Crocco, L.; Isernia, T. Inverse scattering via virtual experiments and contrast source regularization. *IEEE Trans. Antennas Propag.* **2015**, *63*, 1669–1677. [CrossRef]

27. Bevacqua, M.T.; Crocco, L.; Di Donato, L.; Isernia, T. An algebraic solution method for nonlinear inverse scattering. *IEEE Trans. Antennas Propag.* **2015**, *63*, 601–610. [CrossRef]

28. Crocco, L.; Di Donato, L.; Catapano, I.; Isernia, T. An improved simple method for imaging the shape of complex targets. *IEEE Trans. Antennas Propag.* **2013**, *61*, 843–851. [CrossRef]

29. Poli, L.; Oliveri, G.; Rocca, P.; Massa, A. Bayesian compressive sensing approaches for the reconstruction of two-dimensional sparse scatterers under TE illumination. *IEEE Trans. Geosci. Remote Sens.* **2013**, *51*, 2920–2936. [CrossRef]

30. Li, M.; Abubakar, A.; Habashy, T.M. A three-dimensional model-based inversion algorithm using radial basis functions for microwave data. *IEEE Trans. Antennas Propag.* **2012**, *60*, 3361–3372. [CrossRef]

31. Meaney, P.M.; Paulsen, K.D.; Geimer, S.D.; Haider, S.A.; Fanning, M.W. Quantification of 3-D field effects during 2-D microwave imaging. *IEEE Trans. Biomed. Eng.* **2002**, *49*, 708–720. [CrossRef]

32. Semenov, S.Y.; Svenson, R.H.; Bulyshev, A.E.; Souvorov, A.E.; Nazarov, A.G.; Sizov, Y.E.; Pavlovsky, A.V.; Borisov, V.Y.; Voinov, B.A.; Simonova, G.I.; et al. Three-dimensional microwave tomography: Experimental prototype of the system and vector Born reconstruction method. *IEEE Trans. Biomed. Eng.* **1999**, *46*, 937–946. [CrossRef]

33. Bucci, O.M.; Isernia, T. Electromagnetic inverse scattering: retrievable information and measurement strategies. *Radio Sci.* **1997**, *32*, 2123–2137. [CrossRef]

34. Fear, E.C.; Li, X.; Hagness, S.C.; Stuchly, M.A. Confocal microwave imaging for breast cancer detection: Localization of tumors in three dimensions. *IEEE Trans. Biomed. Imaging* **2002**, *49*, 812–822. [CrossRef]

35. Ali, M.A.; Moghaddam, M. 3D nonlinear super-resolution microwave inversion technique using time-domain data. *IEEE Trans. Antennas Propag.* **2010**, *58*, 2327–2336. [CrossRef]

36. Abubakar, A.; Habashy, T.M.; Pan, G.; Li, M. Application of the multiplicative regularized Gauss-Newton algorithm for three-dimensional microwave imaging. *IEEE Trans. Antennas Propag.* **2012**, *60*, 2431–2441. [CrossRef]

37. Donelli, M.; Franceschini, D.; Rocca, P.; Massa, A. Three-dimensional microwave imaging problems solved through an efficient multi-scaling particle swarm optimization. *IEEE Trans. Geosci. Remote Sens.* **2009**, *47*, 1467–1481. [CrossRef]

38. De Zaeytijd, J.; Franchois, A.; Eyraud, C.; Geffrin, J. Full-wave three-dimensional microwave imaging with a regularized Gauss-Newton Method—Theory and experiment. *IEEE Trans. Antennas Propag.* **2007**, *55*, 3279–3292. [CrossRef]

39. Harada, H.; Wall, D.J.N.; Takenake, T.; Tanaka, M. Conjugate gradient method applied to inverse scattering problem. *IEEE Trans. Antennas Propag.* **1995**, *43*, 784–792. [CrossRef]

40. Salucci, M.; Oliveri, G.; Anselmi, N.; Viani, F.; Fedeli, A.; Pastorino, M.; Randazzo, A. Three-dimensional electromagnetic imaging of dielectric targets by means of the multiscaling inexact-Newton method. *J. Opt. Soc. Am. A* **2017**, *34*, 1119–1131. [CrossRef]

41. Estatico, C.; Pastorino, M.; Randazzo, A.; Tavanti, E. Three-Dimensional Microwave Imaging in L^P Banach Spaces: Numerical and Experimental Results. *IEEE Trans. Comput. Imaging* **2018**, *4*, 609–623. [CrossRef]

42. Simonov, N.; Kim, B.; Lee, K.; Jeon, S.; Son, S. Advanced fast 3-D electromagnetic solver for microwave tomography imaging. *IEEE Trans. Med. Imaging* **2017**, *36*, 2160–2170. [CrossRef]

43. Wang, X.Y.; Li, M.; Abubakar, A. Acceleration of 2-D multiplicative regularized contrast source inversion algorithm using paralleled computing architecture. *IEEE Antennas Wirel. Propag. Lett.* **2017**, *16*, 441–444. [CrossRef]

44. Oliveri, G.; Salucci, M.; Anselmi, N.; Massa, A. Compressive sensing as applied to inverse problems for imaging: Theory, applications, current trends, and open challenges. *IEEE Antennas Propag. Mag.* **2017**, *59*, 34–46. [CrossRef]

45. Massa, A.; Rocca, P.; Oliveri, G. Compressive sensing in electromagnetics—A review. *IEEE Antennas Propag. Mag.* **2015**, *57*, 224–238. [CrossRef]

46. Anselmi, N.; Oliveri, G.; Hannan, M.A.; Salucci, M.; Massa, A. Color compressive sensing imaging of arbitrary-shaped scatterers. *IEEE Trans. Microware Theory Technol.* **2017**, *65*, 1986–1999. [CrossRef]

47. Oliveri, G.; Ding, P.-P.; Poli, L. 3D crack detection in anisotropic layered media through a sparseness-regularized solver. *IEEE Antennas Wirel. Propag. Lett.* **2015**, *14*, 1031–1034. [CrossRef]

48. Poli, L.; Oliveri, G.; Viani, F.; Massa, A. MT-BCS-based microwave imaging approach through minimum-norm current expansion. *IEEE Trans. Antennas Propag.* **2013**, *61*, 4722–4732. [CrossRef]

49. Bevacqua, M.T.; Crocco, L.; Di Donato, L.; Isernia, T. Non-linear inverse scattering via sparsity regularized contrast source inversion. *IEEE Trans. Computat. Imag.* **2017**, *3*, 296–304. [CrossRef]

50. Qiu, W.; Zhou, J.; Zhao, H.; Fu, Q. Three-dimensional sparse turntable microwave imaging based on compressive sensing. *IEEE Geosci. Remote Sens. Lett.* **2015**, *12*, 826–830. [CrossRef]

51. Haynes, M.; Stang, J.; Moghaddam, M. Real-time microwave imaging of differential temperature for thermal therapy monitoring. *IEEE Trans. Biomed. Eng.* **2014**, *61*, 1787–1797. [CrossRef]

52. Li, M.; Abubakar, A.; Van Den Berg, P.M. Application of the multiplicative regularized contrast source inversion method on 3D experimental Fresnel data. *Inverse Probl.* **2009**, *25*, 1–23. [CrossRef]

53. Ji, S.; Dunson, D.; Carin, L. Multitask compressive sensing. *IEEE Trans. Signal Process.* **2009**, *57*, 92–106. [CrossRef]

54. Tipping, M.E. Sparse Bayesian learning and the relevant vector machine. *J. Mach. Learn. Res.* **2001**, *1*, 211–244.

55. Bucci, O.M.; Crocco, L.; Isernia, T.; Pascazio, V. Inverse scattering problems with multifrequency data: Reconstruction capabilities and solution strategies. *IEEE Trans. Geosci. Remote Sens.* **2000**, *38*, 1749–1756. [CrossRef]

 Journal of
Imaging

Article

Image-Based RCS Estimation from Near-Field Data

Tushar Rajvanshi [†], Maria Antonia Maisto [†], Angela Dell'Aversano [†] and Raffaele Solimene [*,†]

Department of Engineering, University of Campania, 81031 Aversa, Italy; tushar.rajvanshi@unicampania.it (T.R.); mariaantonia.maisto@unicampania.it (M.A.M.); angela.dellaversano@gmail.com (A.D.)
* Correspondence: raffaele.solimene@unicampania.it; Tel.: +39-081-50-10-335
† These authors contributed equally to this work.

Received: 7 May 2019; Accepted: 14 June 2019; Published: 17 June 2019

Abstract: This paper deals with the problem of estimating the RCS from near-field data by image-based approaches. In particular, a rigorous focusing procedure based on a weighted adjoint scheme, which is also applicable to an arbitrary measurement curve, is developed. The developed formalism allows us to address the important question concerning the need to employ a multi-frequency configuration to estimate the RCS. Accordingly, it is shown that if RCS is required at a given frequency, then the target image obtained solely at such a frequency can be exploited provided that the spatial truncation arising from the size of the investigated area is properly taken into account.

Keywords: RCS estimation; image-based approach; adjoint inversion methods

---

1. Introduction

The Radar Cross Section (RCS) of a target is a crucial quantity that describes how the target responds to an impinging electromagnetic wave (i.e., how it scatters the incident wave) across different directions. As is well known, RCS is formally defined as the distance between the probing antenna and the target approaching infinity [1]. In practical cases, to measure the RCS, the physical separation between the target and the illuminating/measuring antenna is required to comply with the far-field conditions. However, as frequency increases, and depending on the size of the target, the required separation can soon become very large. Therefore, for a long time, there has been a great interest in developing methods for predicting the RCS for scattered field measurements taken at a distance not in far-field [2–6]. Indeed, the possibility of acquiring data in near-zone avoids the above-mentioned drawback and in principle offers several advantages since measurements can be taken within an anechoic chamber. On the other hand, near-zone measurements cannot be used straight away for RCS estimation, due to the wavefront curvature and because in near-zone the target is not uniformly illuminated. Compact range equipment solves these problems, but requires high-quality reflectors [4,6]. Alternatively, image-based approaches first obtain a reconstruction of the target reflectivity and then the RCS is estimated by Fourier, transforming the obtained reflectivity. Accordingly, those approaches can be regarded as particular near-field to far-field transformations [7]. One can consider compact range methods and image-based approaches to move the job to do from hardware, i.e., the reflector, to software, i.e., the scattered field data processing algorithms.

In this paper, we are concerned with image-based approaches, in particular under a monostatic measurement setup.

The underlying working assumption is that the target scatters linearly, i.e., multiple scattering, shadowing, and creeping waves are considered negligible. Accordingly, the scattering phenomenon can be described by a linear integral scattering operator and the reconstruction of the target reflectivity is usually obtained by solving (inverting) that integral equation via some focusing procedure borrowed from back-propagation/migration algorithms literature [8].

Focusing entails some smearing in the reconstruction (depending on the measurement configuration parameters such as frequency band, etc.) which can negatively impact the subsequent RCS estimation. To compensate for such an effect, the obtained RCS can be normalized by the one corresponding to a point-like target [9] or, alternatively, the standard focusing kernel can be corrected by a suitable factor [10].

The first procedure implicitly assumes a spatially invariant behavior (within the image region) of the imaging procedure which indeed does not hold. The second approach has instead been demonstrated only for a circular measurement curve. The first contribution of this paper is the derivation of the focusing procedure under a general framework as far as the measurement curve is concerned, which actually generalizes the procedure reported in [10] to a generally shaped measurement curve.

To obtain high-resolution target reconstructions, the imaging stage is usually achieved by using multi-frequency data. However, target reflectivity in general depends on frequency. As outlined in [7], in that case, the obtained reflectivity can be considered more like an average over the employed frequency band. Furthermore, the obtained reflectivity reconstruction is generally used for RCS estimation just at the central frequency. It is then natural to ask if the imaging procedure can work by directly using single-frequency data.

This is indeed possible [11] even though RCS estimation is more sensitive to the spatial truncation determined by the size of the image region.

2. Basics on Image-Based RCS

In this section, we introduce the problem, the adopted notation, and briefly recall image-based methods present in literature.

The scattering experiment related to the RCS estimation considered in this paper is described in Figure 1. In particular, we refer to a 2D scalar configuration, i.e., the target is invariant along the z-axis and the probing field is a cylindrical wave linearly polarized again along z. Accordingly, the scattered field is also linearly polarized and the RCS is a single scalar and not a dyad, as in the general 3D vector case. The target is assumed to reside within the *image* domain D_I, whereas the scattered field is collected in near-field over a curve Γ surrounding D_I and for the frequency band Ω. The usual monostatic configuration is considered. Accordingly, during the data-acquisition stage, the same antenna acts as transmitter and receiver and then moves around the target to synthesize the measurement curve, or equivalently the target is rotated over a turntable. In the latter case the synthesized measurement curve is just a circle, which is the one that is usually considered in literature.

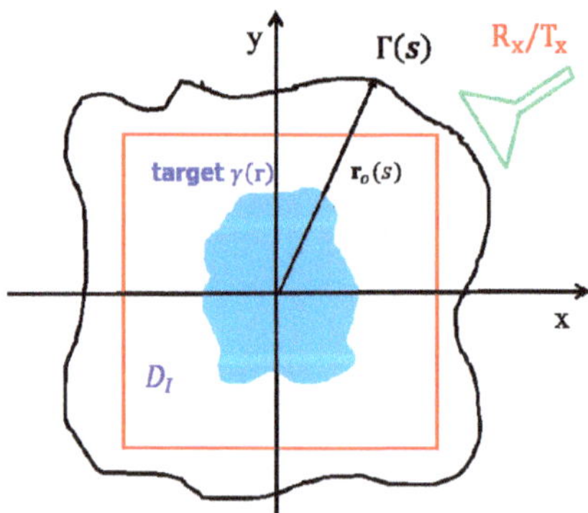

Figure 1. Geometry of the problem.

It is worth remarking that we are considering such a 2D configuration for the sake of simplicity, since this allows easier production of the numerical examples to be shown later on. However, a similar configuration is commonly addressed also in 3D cases while estimating the so-called *water line* RCS [7].

The starting point for problem formulation is the assumption that the target consists of an ensemble of independent and non-directional scattering centers [9]. Since multiple interactions, creeping waves and shadowing effects are neglected, the target reflectivity $\gamma(\mathbf{r})$ and the scattered field are linked by the following linear integral operator

$$E_S(k, \mathbf{r}_o(s)) = \int_{D_I} A(k, \mathbf{r}_o(s), \mathbf{r}) e^{-j\phi[k, \mathbf{r}_o(s), \mathbf{r}]} \gamma(\mathbf{r}) d^2\mathbf{r} \tag{1}$$

where k is the wavenumber and $\mathbf{r}_o(s) \in \Gamma$ is the measurement curve parametrized with respect to s. Moreover, the phase term takes into account the propagation path from the antenna to the target and back, i.e.,

$$\phi(k, \mathbf{r}_o, \mathbf{r}) = 2k|\mathbf{r}_o(s) - \mathbf{r}|$$

whereas the amplitude term is given by

$$A(k, \mathbf{r}_o(s), \mathbf{r}) = \frac{\alpha C(k) P^2[\theta(\mathbf{r}_o(s), \mathbf{r})]}{|\mathbf{r}_o(s) - \mathbf{r}|}$$

with α being a constant term that varies depending on whether a 3D or 2D geometry is being considered, $P[\theta(\mathbf{r}_o(s), \mathbf{r})]$ is the antenna directivity pattern assumed to be 1 for $\theta(\mathbf{r}_o(s), \mathbf{r}) = 0$, (with $\theta(\mathbf{r}_o(s), \mathbf{r})$ being the angle between the $\mathbf{r}_o(s) - \mathbf{r}$ and the antenna broadside direction) and $C(k)$ collects the antenna frequency behavior and the frequency part of the propagator (Green function) amplitude. In particular, for the case at hand, the relevant Green function is proportional to a Hankel function of zero order and second kind. $P[\theta(\mathbf{r}_o(s), \mathbf{r})]$ and $C(k)$ are assumed to be known by a preliminary stage of measurement and calibration using a reference target of known RCS such as a metallic sphere, a plate or the like.

At this juncture, some further considerations concerning the model (1) are in order. First, it is noted that in (1) each single point in the image domain has been considered to be being in the far-field of the transmitting/receiving antenna. If this is not the case, a linear transformation could be still established by invoking the plane-wave spectrum representation for both the Green function and the antenna response. This circumstance is not considered in this manuscript. Second, field data, instead of voltage, are being considered. Voltage is actually what one can measure. However, assuming field data does not impair the generality of the theoretical/numerical analysis we intend to pursue. Finally, the target reflectivity is in general frequency dependent; we denote such a dependence by $\gamma_f(k)$. However, if the target can be represented by an ensemble of point-like scatterers, the frequency dependent part of reflectivity is a priori known and hence can be considered embodied within the $C(k)$ term [10]. However, we will turn back on this assumption in the following sections.

The near-field (1) is in general much different from the corresponding far-field and hence cannot be used directly for estimating the target RCS. This of course is due to the wavefront curvature of the impinging electromagnetic wave and to the non-uniform illumination of the target. The aim of the imaging stage is just to compensate for such wavefront curvature and amplitude behavior. This is achieved by processing the scattered field data through a focusing operator that actually "translates" the scattered field data into an image of the target under test. Formally, this is written as

$$\hat{\gamma}(\mathbf{r}) = \int_{\Omega \times \Gamma} f_c[k, \mathbf{r}_o(s), \mathbf{r}] E_S[k, \mathbf{r}_o(s)] dk ds \tag{2}$$

where $f_c[k, \mathbf{r}_o(s), \mathbf{r}]$ is the focusing kernel which is commonly chosen equal to $e^{j\phi[k, \mathbf{r}_o(s), \mathbf{r}]} / A[k, \mathbf{r}_o(s), \mathbf{r}]$. Clearly, $\hat{\gamma}$ is a filtered (blurred) version of the actual reflectivity. That filtering depends on the configuration parameters, i.e., $\Omega \times \Gamma$, which reflect the properties of the imaging point-spread function, i.e.,

$$psf(\mathbf{r}, \mathbf{r}') = \int_{\Omega \times \Gamma} \frac{A[k, \mathbf{r}_0(s), \mathbf{r}']}{A[k, \mathbf{r}_0(s), \mathbf{r}]} e^{j[\phi[k, \mathbf{r}_0(s), \mathbf{r}] - \phi[k, \mathbf{r}_0(s), \mathbf{r}']]} dk ds \tag{3}$$

so that (2) can be equivalently rewritten as

$$\tilde{\gamma}(\mathbf{r}) = \int_{D_I} psf(\mathbf{r}, \mathbf{r}')\gamma(\mathbf{r}')d^2\mathbf{r}' \tag{4}$$

Once the reflectivity has been estimated, the RCS can be computed by using its (2D) definition equation

$$\sigma(k, \theta_o) = \lim_{r_o \to \infty} 2\pi r_o \left| \frac{E_S(k, \mathbf{r}_o)}{E_{inc}} \right|^2 \tag{5}$$

with $\mathbf{r}_o \equiv (r_o, \theta_o)$ being the observation point. Hence, on exploiting (1) and (2), (5) yields

$$\sigma(k, \theta_o) = \left| \alpha\gamma_f(k) \int_{D_I} \tilde{\gamma}(\mathbf{r})e^{2jk\hat{\mathbf{r}}_o \cdot \mathbf{r}} d^2\mathbf{r} \right|^2 \tag{6}$$

Equation (6) allows recognition that under the assumed linear scattering model, the RCS depends of the Fourier transform of the reflectivity function projected over the so-called Ewald disc (resp. sphere for the 3D case) [12]. Of course, because of the blurring introduced by the imaging procedure, some error will corrupt the estimation (6). In order to mitigate such an error, a common way to proceed is to normalize (6) by the RCS of the point-spread function [9], which is computed by Fourier transforming $psf(\mathbf{r}, 0)$. It is clear that behind this procedure there is the implicit assumption of considering (4) as a convolution which in general does not hold. Differently, in [10], it is shown that by introducing a suitable correction term, the focusing procedure can be approximated as a Fourier transformation. Accordingly, $\tilde{\gamma}$ proves to be a windowed (in spatial spectrum domain) Fourier transform of the γ; therefore, normalization is no longer required.

We will come back to these important points in the next section where we introduce an imaging procedure for a more general (with respect to the usual circle) measurement curve.

3. Adjoint Inversion for Generic Measurement Curve

According to the previous section, the first step in any image-based RCS estimation method is to solve the integral Equation (1) for the reflectivity γ. Formally, this entails finding the inverse of the linearized scattering operator

$$\mathcal{A}_S : \gamma \to E_S \tag{7}$$

$\mathcal{A}_S$ being just the integral operator in (1). A very common way to invert (7) is to adopt the adjoint operator $\mathcal{A}_S^\dagger$ instead of the inverse of $\mathcal{A}_S$. A number of popular methods such as time-reversal, reverse-migration, back-propagation, etc. are adjoint-based imaging schemes [8]. Inversion through the adjoint allows us to deal with the ill-posedness of the problem in that it retunes a stable reconstruction procedure. However, it is not a Tichonov regularization scheme as the corresponding reconstruction fails to converge to the generalized solution even in absence of noise [8]. Their great diffusion is due to their simple *physical* understanding and the possibility to be often implemented via FFT. In particular, adjoint inversion succeeds in compensating the phase in correspondence to the actual scatterers's position but the amplitude is not addressed properly. For this reason, adjoint inversion is often paired with a pre-weighting (filtering) stage, so that the image/reconstruction is obtained as [13]

$$\mathcal{A}_S^\dagger : WE_S \to \tilde{\gamma} \tag{8}$$

where $W(k, \mathbf{r}_0, \mathbf{r})$ is just the weighting function. In particular, on exploiting (1), the reconstruction operator (8) can be explicitly written as

$$\tilde{\gamma}(\mathbf{r}) = \int_{\Omega \times \Gamma} W[k, \mathbf{r}_0(s), \mathbf{r}] e^{j\phi[k, \mathbf{r}_0(s), \mathbf{r}]} E_S[k, \mathbf{r}_0(s)] dk ds \tag{9}$$

which yields the point-spread function

$$psf(\mathbf{r}, \mathbf{r}') = \int_{\Omega \times \Gamma} W[k, \mathbf{r}_0(s), \mathbf{r}] A[k, \mathbf{r}_0(s), \mathbf{r}'] e^{j\{\phi[k, \mathbf{r}_0(s), \mathbf{r}] - \phi[k, \mathbf{r}_0(s), \mathbf{r}']\}} dk ds \tag{10}$$

Please note that the usual choice for the weighting function, as done in the previous section, is to $W[k, \mathbf{r}_0(s), \mathbf{r}] = 1 / A[k, \mathbf{r}_0(s), \mathbf{r}]$. On the other hand, it is desirable that the point-spread function be as close as possible to a delta function. If this is achieved, the adjoint inversion actually approximates a regularized reconstruction [14]. To cope with this question, the weighting function must be properly chosen as detailed below.

We start by introducing the variable

$$\mathbf{w}[k, \mathbf{r}_0(s), \mathbf{r}, \mathbf{r}'] = - \int_0^1 \nabla_\mathbf{x} \phi[k, \mathbf{r}_0(s), \mathbf{x}]|_{\mathbf{x} = \mathbf{r}' + \lambda(\mathbf{r} - \mathbf{r}')} d\lambda \tag{11}$$

which allows rewriting of the phase term in (10) as

$$\phi[k, \mathbf{r}_0(s), \mathbf{r}] - \phi[k, \mathbf{r}_0(s), \mathbf{r}'] = -\mathbf{w}[k, \mathbf{r}_0(s), \mathbf{r}, \mathbf{r}'] \cdot (\mathbf{r} - \mathbf{r}') \tag{12}$$

and the point-spread function (10) as

$$psf(\mathbf{r}, \mathbf{r}') = \int_{\Omega_\mathbf{w}} W(\mathbf{w}, \mathbf{r}) A(\mathbf{w}, \mathbf{r}') e^{-j\mathbf{w} \cdot (\mathbf{r} - \mathbf{r}')} |\mathbf{J}| d^2 \mathbf{w} \tag{13}$$

with $\mathbf{J}$ being the Jacobian of the transformation that maps $\mathbf{w}$ in (k, s). Equation (13) shows the point-spread function as pseudodifferential operators which enjoy the so-called pseudolocal property [15]. This is the very mathematical rationale that justifies performing the adjoint inversion to retrieve object singularities, even though in order to correctly retrieve the singularity amplitudes the weighting function must be properly chosen. Also, (13) makes it immediately clear that the leading order contribution occurs for $\mathbf{r} - \mathbf{r}' = 0$. Accordingly, the approximation

$$\mathbf{w}[k, \mathbf{r}_0(s), \mathbf{r}, \mathbf{r}'] = \mathbf{w}[k, \mathbf{r}_0(s), \mathbf{r}, \mathbf{r}] = -\nabla_\mathbf{x} \phi[k, \mathbf{r}_0(s), \mathbf{x}]|_{\mathbf{x} = \mathbf{r}} \tag{14}$$

is made in (13). This leads to the following point-spread function approximation

$$psf(\mathbf{r}, \mathbf{r}') \simeq \int_{\Omega_\mathbf{w}} e^{-j\mathbf{w} \cdot (\mathbf{r} - \mathbf{r}')} d^2 \mathbf{w} \tag{15}$$

once the weighting function is chosen as

$$W(\mathbf{w}, \mathbf{r}) = 1 / [A(\mathbf{w}, \mathbf{r}) |\mathbf{J}|] \tag{16}$$

It is seen that (15) is a filtered version of a Dirac delta. Moreover, Equation (15) makes it clear the spectral content of the target that can be retrieved. Indeed, when the observation curve Γ goes around the target it results that

$$\Omega_\mathbf{w} = \{\mathbf{w} : 2k_{min} \leq w \leq 2k_{max}\} \tag{17}$$

where k_{min} and k_{max} are the wavenumbers corresponding to the lowest and highest adopted frequencies. This result could have been expected since it exactly coincides to what can be retrieved by a far-field configuration. However, it must be kept in mind that (17), and of course (15), holds only approximately true because of the approximation (14). However, (14) is expected to work well in reproducing the main beam of the point-spread function, which in turn plays the major role in the filtering introduced by the imaging procedure.

It is interesting to highlight that (15) holds true regardless of the shape of the observation curve Γ. Indeed, the shape of Γ only enters in the choice of the weighting function through the Jacobian term. In particular, when Γ is a circle surrounding the image region (as it is commonly assumed) simple calculations show that (15) exactly returns the imaging kernel introduced in [10], with the correcting term there introduced just being given by $1/|\mathbf{J}|$. In this regard, the formulation introduced in this paper generalizes previous literature results, which indeed have mainly considered circular measurement curves. Previous discussion can be summarized by the following statement: the resolution achievable during the image stage does not depend on the shape of the measurement curve. Moreover, by exploiting techniques similar to those in [16], an analytical expression for the point-spread function can be obtained. In particular, it can be shown that the point-spread function in (15) is given as

$$psf(\mathbf{r}, \mathbf{r}') \simeq \frac{2\pi[\psi(2k_{max}|\mathbf{r} - \mathbf{r}'|) - \psi(2k_{min}|\mathbf{r} - \mathbf{r}'|)]}{|\mathbf{r} - \mathbf{r}'|^2} \tag{18}$$

with

$$\psi(x) = \int_0^x y J_0(y) dy = x J_1(x) \tag{19}$$

and $J_0(\cdot)$ and $J_1(\cdot)$ being Bessel functions of zero and first order. In Figure 2, the comparison between the actual point-spread function and the one returned by (18) is shown. As can be appreciated, the two curves overlap very well, this means that the leading term approximation exploited to derive (18) works fine.

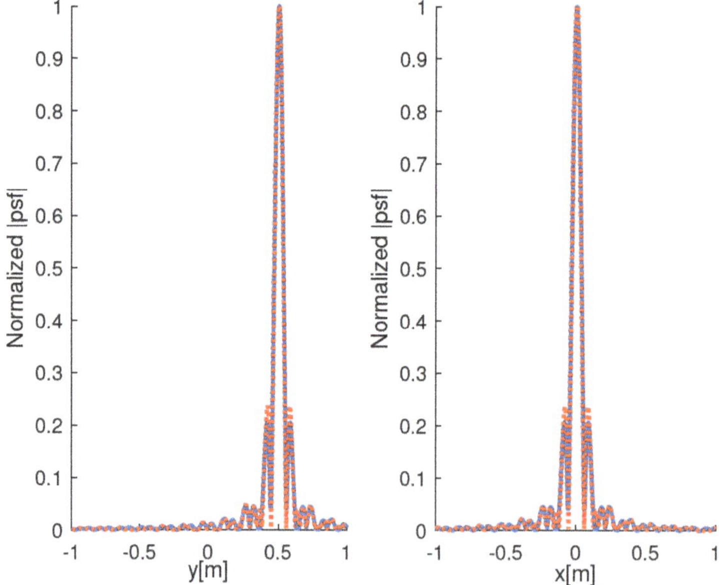

Figure 2. Comparison between the normalized point-spread function amplitude (blue line) and the one returned by (18) (red dotted line). The left panel refers to the cut along the y-axis; the right one to the cut along the x-axis. The adopted frequency band is $\Omega = [0.5, 1.5]$ GHz while the observation curve is an ellipse with axes of 2 m and 2.5 m, respectively.

A further point that is worth stressing is that for a full-view scan (i.e., for a measurement curve running all around the image region) the spectral domain $\Omega_\mathbf{w}$ does not depend on the image point $\mathbf{r}$. This means that the image kernel (15) can be actually considered to be convolution; however, this does not hold true for (10), since to get (15) a spatially varying kernel was required. Accordingly, the normalization procedure in [9] can be expected to work only when the image points are close to the reference one (i.e., the one for which the point-spread function is computed and transformed).

Finally, it is remarked that to speed up the imaging procedure FFT-based routines are often employed [17]. According to the previous formulation, this can be achieved for a generic measurement curve. Of course, the usual interpolation and resampling step, which is required to map data over a uniform grid, in general depends on the particular curve under concern. In the following numerical examples we do not follow such a procedure, rather we employ (9) with the weighting function chosen as in (16).

4. Single-Frequency Imaging

As clearly stated above, the previous RCS estimation procedure is founded on the assumption that the targets are frequency independent or their frequency behaviors are equal and known. The latter holds true for point-like targets but for most practical cases, even when targets can be still described by an ensemble of scattering centers, the different contributions depend on the operating frequency. Accordingly, the reflectivity returned by (9) is actually an averaged version performed over the frequency band. This circumstance in general impacts negatively on the RCS computation. That is why, the obtained image is usually used to compute the RCS only at the central (of the adopted band) frequency [7].

Let us relax the above-mentioned assumption. Hence, here we consider the cases in which the target frequency behavior is unknown, or it consists of scattering centers whose reflectivity coefficients differently depends on frequency. In these cases, the frequency behavior cannot be singled out and embodied within the scattering operator since actually $\gamma = \gamma(k, \mathbf{r})$. Accordingly, the scattered field is yielded by

$$E_S[k, \mathbf{r}_o(s)] = \int_{D_I} A[k, \mathbf{r}_o(s), \mathbf{r}] e^{-j\phi[k, \mathbf{r}_o(s), \mathbf{r}]} \gamma(k, \mathbf{r}) d^2\mathbf{r} \tag{20}$$

and the estimated reflectivity (through imaging) and the actual one are linked as

$$\tilde{\gamma}(\mathbf{r}) = \int_\Omega dk \overbrace{\int_{D_I} \int_\Gamma f_c[k, \mathbf{r}_o(s), \mathbf{r}] A[k, \mathbf{r}_o(s), \mathbf{r}'] e^{-j\phi[k, \mathbf{r}_o(s), \mathbf{r}']} \gamma(k, \mathbf{r}')}^{\tilde{\gamma}(k, \mathbf{r})} ds d^2\mathbf{r}' \tag{21}$$

It is seen that $\tilde{\gamma}(\mathbf{r})$ is now given as the coherent summation of the single-frequency images $\tilde{\gamma}(k, \mathbf{r})$ and as such it does not represent the (regularized) solution of (20). This could have been expected since both scattered field data and unknown reflectivity depend on the frequency. As discussed above, this leads to a degradation of accuracy while estimating the RCS.

The arising question is whether, in order to estimate the RCS at a given frequency, a frequency band must necessarily be employed. The answer to this question seems positive if high-resolution radar imaging is aimed. At the other hand, single-frequency images are not affected by previous drawback but the performance in the reconstruction in general results much lower. Hence, the question can be rephrased as whether image degradation (due to single-frequency data) impairs RCS estimation. To this end, we particularize (15) to the single-frequency case, for example by considering $k = k_{av}$ (i.e., the average one)

$$psf_{sf}(\mathbf{r}, \mathbf{r}') \simeq \int_{C_\mathbf{w}} e^{-j\mathbf{w} \cdot (\mathbf{r} - \mathbf{r}')} d^2\mathbf{w} \tag{22}$$

where the subscript sf stands for single-frequency and $C_\mathbf{w} = \{\mathbf{w} : w = 2k\}$. (22) can be easily computed and the result is

$$psf_{sf}(\mathbf{r}, \mathbf{r}') \simeq 4\pi k J_0(2k|\mathbf{r} - \mathbf{r}'|) \tag{23}$$

In Figure 3 the comparison between the actual point-spread function and the one returned by (23) is shown. Again, the estimated psf allows us to obtain a good approximation of the actual one.

To understand how resolution degrades when single-frequency data are used, Equation (23) can be compared to (18). It is clear that now the spatial spectrum that can be retrieved about the unknown reflectivity is much lower since $C_\mathbf{w}$ consists of a single circle instead of the circular annulus $\Omega_\mathbf{w}$. Nonetheless, as shown in [16], it is the relative bandwidth (i.e., $(k_{max} - k_{min})/k_{av}$), rather than the frequency band, that plays the major role. This is somewhat different from the common belief that the frequency band is necessary and is a consequence of the full-view (i.e., measurements are taken all around the target) configuration. However, we do not want to dwell any further on that point since this is not the focus of the paper. The point is whether image degradation impairs the possibility of obtaining RCS. To this end, we once again remark that (22) introduces a filtering that allows retention of only the target reflectivity over the circle $C_\mathbf{w}$, which is the Ewald circle corresponding to the wavenumber k and which is, according to (6), what is necessary to compute the RCS.

Finally, by looking at Figure 3, it can be seen that the side-lobes are higher than the multi-frequency case. This allows expectation that while comparing multi-frequency and single-frequency RCS estimations, the spatial truncation introduced by the size of the investigated area will play a crucial role.

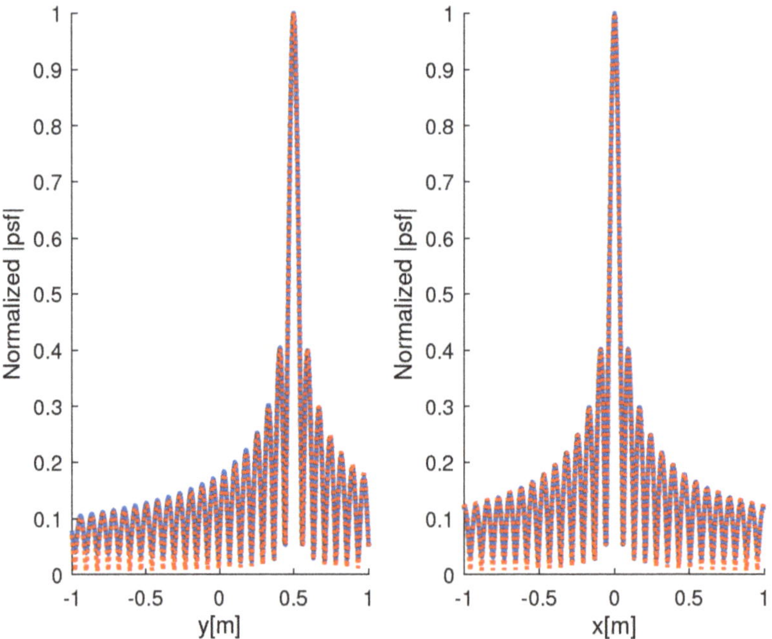

Figure 3. Comparison between the normalized point-spread function amplitude (blue line) and the one returned by (23) (red dotted line). The left panel refers to the cut along the y-axis whereas the right one to the cut along the x-axis. The frequency is 1 GHz while the observation curve is an ellipse with axes 2 m and 2.5 m.

To check previous arguments, we conclude this section by running a simple case. In particular, we consider two point-like objects whose reflectivity is given by $\gamma(x,y) = 0.1(j\frac{k}{K_{min}})^2 [\delta(x, y - 0.5) + \delta(x, y + 0.5)]$. The electric field is collected over an ellipse whose axes are 3 m and 3.5 m and the working frequency is $f = 1.75$ GHz. Since the far-field distance is 11.7 m, the ellipse lies in the near-field region of two point scatterers. In Figure 4 the RCS at that frequency is shown. The green lines refer to the actual RCS, while the blue and red lines to the ones computed from the multi-frequency ($\Omega = [0.5, 3]\,GHz$) and single-frequency images, respectively. In particular, in the panel a) $D_I = [(-1, 1) \times (-1, 1)]$ is considered whereas for panel b) $D_I = [(-2, 2) \times (-2, 2)]$. From the Figure 4a) it can be appreciated that the multi-frequency configuration allows us to obtain a better RCS estimation. However, when D_I increases (see Figure 4b) the red line becomes very similar to the blue one. This confirms that single-frequency RCS is feasible and competitive with respect to the multi-frequency one provided the size of the investigated area is enlarged.

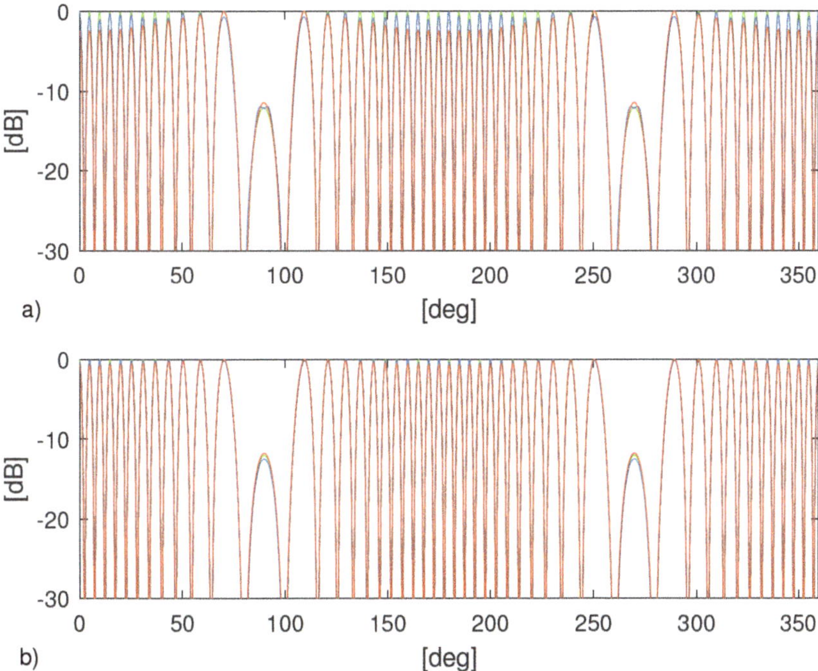

Figure 4. Normalized RCS (in db scale) at $f = 1.75$ GHz. The green lines refer to the actual RCS, while the blue and red lines to the ones computed from the multi-frequency and single-frequency images, respectively. (**a**) $D_I = [(-1, 1) \times (-1, 1)]$ and (**b**) $D_I = [(-2, 2) \times (-2, 2)]$.

5. Numerical Examples

In this section, some further examples are addressed to check RCS estimation.

In all the following examples, the RCS is evaluated at $f_{avg} = 4.5$ GHz, the field is collected over an ellipse whose axes are $30\lambda_{avg}$ and $35\lambda_{avg}$ with λ_{avg} the wavelength at f_{avg} and $D_I = [(-20\lambda_{avg}, 20\lambda_{avg}) \times (-20\lambda_{avg}, 20\lambda_{avg})]$. Moreover, when the multi-frequency configuration is exploited $\Omega = [3, 6]$ GHz. The target, always a perfect electric conducting scatterer, is contained within a circular image region with radius $5\sqrt{2}\lambda_{avg}$. Accordingly, the far-field distance is $400\lambda_{avg}$ and the ellipse lies in the near-field region of the object. Three different objects, shown in Figure 5, are considered.

The first object is a bar of length $10\sqrt{2}\lambda_{avg}$. The corresponding multi-frequency and single-frequency images are reported in the panels (a) and (b) of Figure 6, while the RCS estimations are compared in (c). As can be seen, despite the evident degradation of the reconstructed image, single-frequency configuration allows us to obtain an RCS estimation which is very similar to the multi-frequency one; indeed the three curves overlap very well.

The circular object (of radius $5\sqrt{2}\lambda_{avg}$) is addressed in Figure 7. As can be seen, the multi-frequency reconstruction allows us to clearly discern the shape of the target which is completely lost in the single-frequency case. Nonetheless. the RCS estimation is comparable with the multi-frequency one.

The last considered example, the missile shaped target is addressed in Figure 8. The same discussion as for the previous examples still holds. In this case, it must be remarked that even though the main RCS features (spikes) are captured, both the multi-frequency and the single-frequency approaches show some deviation from the actual RCS. This can be ascribed to the model error (for example multiple scattering, which is more relevant for this complex shaped target) that inherently affects image-based procedure.

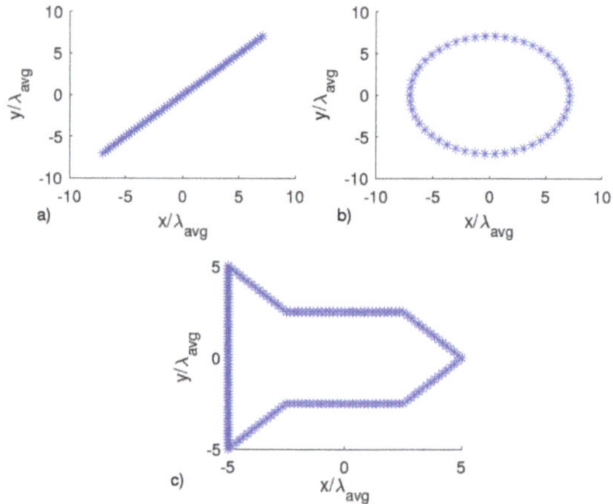

Figure 5. Contours of the objects under test.

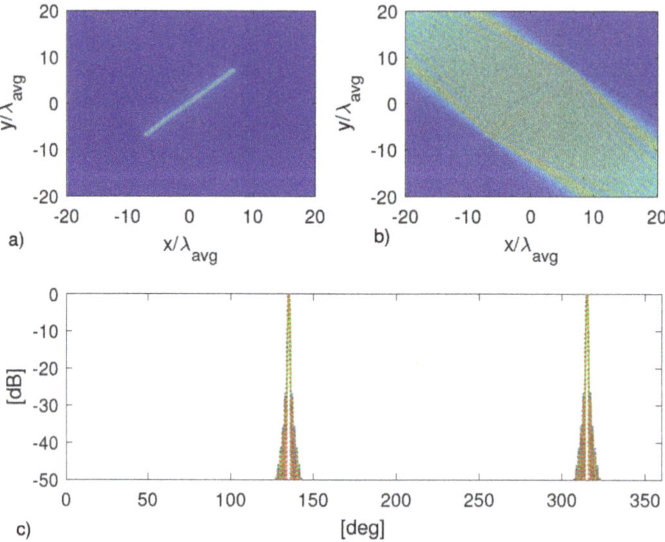

Figure 6. Object in Figure 5a. (**a**) Multi-frequency image. (**b**) Single-frequency image. (**c**) Normalized RCS at $f = 4.5\,\text{GHz}$. The green line refers to the actual RCS, while the blue and red lines to the ones computed from multi-frequency and single-frequency images, respectively.

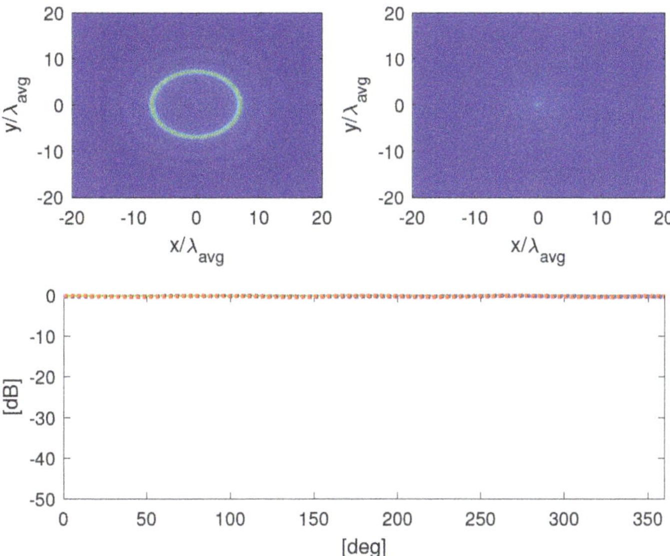

Figure 7. Object in Figure 5b. (**a**) Multi-frequency image. (**b**) Single-frequency image. (**c**) Normalized RCS at $f = 4.5$ GHz. The green line refers to the actual RCS, while the blue and red lines to the ones computed from the multi-frequency and single-frequency images, respectively.

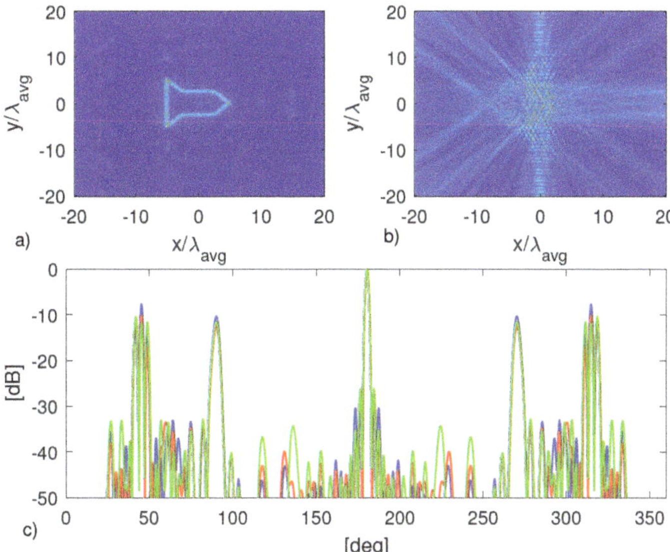

Figure 8. Object in Figure 5c. (**a**) Multi-frequency image. (**b**) Single-frequency image. (**c**) Normalized RCS at $f = 4.5$ GHz. The green line refer to the actual RCS, while the blue and red lines to the ones computed from the multi-frequency and single-frequency images, respectively.

6. Conclusions

In this paper, the problem of estimating the RCS from near-field data by image-based approaches has been addressed.

The first contribution we conveyed in this manuscript was the rigorous derivation of a focusing procedure based on a weighted adjoint scheme which generalizes for an arbitrary measurement curve the results presented in [10] that have been developed for a circular measurement curve.

Secondly, the important question concerning the necessity to use a multi-frequency configuration to estimate the RCS has been studied. Accordingly, a deep analysis of the introduced analytical tools highlighted that if RCS is required at a given frequency, then the target image obtained solely at such a frequency can be in principle exploited. This opens the possibility of employing cheaper measurement systems and to take into account target frequency dependence. However, the spatial truncation introduced by the size of the investigated area must be properly taken into account.

Several numerical examples corroborated the possibility of computing the RCS from single-frequency image.

For the sake of simplicity, the study has been developed for a 2D scalar configuration. The generalization to 3D real scenario is a possible future development.

Author Contributions: Conceptualization, R.S. and M.A.M.; methodology, R.S. and M.A.M.; software, T.R and A.D.; validation, T.R. and A.D.; formal analysis, R.S.; investigation, M.A.M.; resources, R.S., M.A.M., T.R. and A.D.; data curation, R.S., M.A.M., T.R. and A.D.; writing–original draft preparation, R.S., M.A.M., T.R. and A.D.; writing–review and editing, R.S., M.A.M., T.R. and A.D.; visualization, R.S., M.A.M., T.R. and A.D.; supervision, R.S.; project administration, R.S.; funding acquisition, R.S.

Funding: This research received no external funding.

Acknowledgments: The authors kindly thank Giuseppina Nuzzo for proofreading the manuscript.

Conflicts of Interest: The authors declare no conflict of interest.

References

1. Balanis, A.C. *Antenna Theory: Analysis and Design*, 3rd ed.; John Wiley and Sons: Hoboken, NJ, USA, 2005.

2. Falconer, D.G. Extrapolation of Near-Field RCS Measurements to the Far Zone. *IEEE Trans. Antennas Propag.* **1988**, *36*, 822–829. [CrossRef]

3. Cown, B.J.; Ryan, C.E. Near-field scattering measurements for determining complex target RCS. *IEEE Trans. Antennas Propag.* **1989**, *37*, 576–585. [CrossRef]

4. Theron, I.P.; Walton, E.K.; Gunawan, S. Compact range radar cross-section measurements using a noise radar. *IEEE Trans. Antennas Propag.* **1998**, *46*, 1285–1288. [CrossRef]

5. LaHaie, I.J. Overview of an image-based technique for predicting far-field radar cross section form near-field measurements. *IEEE Antennas Propag. Mag.* **2003**, *45*, 159–169. [CrossRef]

6. Sevgi, L.; Rafiq, Z.; Majid, I. Radar Cross Section (RCS) Measurements. *IEEE Antennas Propag. Mag.* **2013**, *55*, 277–291. [CrossRef]

7. Odendaal, J.W.; Joubert, J. Radar cross section measurements using near-field radar imaging. *IEEE Trans. Instrum. Meas.* **1996**, *45*, 948–954. [CrossRef]

8. Solimene, R.; Catapano, I.; Gennarelli, G.; Cuccaro, A.; Dell'Aversano, A.; Soldovieri, F. SAR imaging algorithms and some unconventional applications: A unified mathematical overview. *IEEE Signal Process. Mag.* **2014**, *31*, 90–98. [CrossRef]

9. Broquetas, A.; Palau, J.; Jofre, L.; Cardama, A. Spherical wave near-field imaging and radar cross-section measurement. *IEEE Trans. Antennas Propag.* **1998**, *46*, 730–735. [CrossRef]

10. Osipov, A.; Kobayashi, H.; Suzuki, H. An Improved Image-Based Circular Near-Field-to-Far-Field Transformation. *IEEE Trans. Antennas Propag.* **2013**, *61*, 989–993. [CrossRef]

11. Schnattinger, G.; Mauermayer, R.A.M.; Eibert, T.F. Monostatic Radar Cross Section Near-Field Far-Field Transformations by Multilevel Plane-Wave Decomposition. *IEEE Trans. Antennas Propag.* **2014**, *62*, 4259–4268. [CrossRef]

12. Devaney, A.J. *Mathematical Fundations of Imaging Tomography and Wavefield Inversion*; Cambridge University Press: New York, NY, USA, 2012.
13. Solimene, R.; Cuccaro, A. Back-propagation imaging by exploiting multipath from point scatterers. *Inverse Probl.* **2017**, *33*, 105010. [CrossRef]
14. Cuccaro, A.; Solimene, R. Inverse Source Problem for a Host Medium Having Pointlike Inhomogeneities. *IEEE Trans. Geosci. Remote Sens.* **2018**, *56*, 5148–5159. [CrossRef]
15. Cheney, M.; Bonneau, R.J. Imaging that exploits multipath scattering from point scatterers. *Inverse Probl.* **2004**, *30*, 1691–1711. [CrossRef]
16. Solimene, R.; Cuccaro, A.; Ruvio, G.; Tapia, D.F.; O'Halloran, M. Beamforming and holography image formation methods: An analytic study. *Opt. Express* **2016**,*8*, 9077–9093. [CrossRef] [PubMed]
17. Vaupel, T.; Eibert, F. Comparison and Application of Near-Field ISAR Imaging Techniques for Far-Field Radar Cross Section Determination. *IEEE Trans. Antennas Propag.* **2006**, *54*, 144–151. [CrossRef]

Article

A Minimum Rank Approach for Reduction of Environmental Noise in Near-Field Array Antenna Diagnosis

Marco Donald Migliore *,†, **Fulvio Schettino** †, **Daniele Pinchera** †, **Mario Lucido** † and **Gaetano Panariello** †

Department of Electrical and Information Engineering, University of Cassino and Southern Lazio, Via Di Bisaio 43, 03043 Cassino, Italy; schettinoo@unicas.it (F.S.); pinchera@unicas.it (D.P.); lucido@unicas.it (M.L.); panariello@unicas.it (G.P.)
* Correspondence: mdmiglio@unicas.it; Tel.: +39-0776-299-3550
† Current address: DIEI, Via Di Biasio 43, 03043 Cassino, Italy.

Received: 5 February 2019; Accepted: 28 April 2019; Published: 2 May 2019

Abstract: A method to filter out the contribution of interference sources in array diagnosis is proposed. The interference-affected near field measured on a surface is treated as a (complex-data) image. This allows to use some modern image processing algorithms. In particular, two strategies widely used in image processing are applied. The first one is the reduction of the amount of information by acquiring only the innovation part of an image, as currently happens in video processing. More specifically, a differential measurement technique is used to formulate the estimation of the array excitations as a sparse recovery problem. The second technique has been recently proposed in video denoising, where the image is split into a low-rank and high-rank part. In particular, in this paper the interference field is filtered out using sparsity as discriminant adopting a mixed minimum ℓ_1 norm and trace norm minimization algorithm. The methodology can be applied to both near and far field measurement ranges. It could be an alternative to the systematic use of anechoic chambers for antenna array testing.

Keywords: antenna array; near-field measurements; 5G communication; array diagnosis; rank minimization; compressed sensing; antenna testing

---

1. Introduction

Sophisticated radiating systems as active array antennas, and massive MIMO arrays will play a relevant role in the forthcoming 5G communication systems [1]. Due to the high levels of electronic devices integration required in 5G antennas, no physical connectors are generally available. This yields a radically new connectorless measurement paradigm in which over-the-air (OTA) measurements will have a relevant role. Furthermore, mass production of these new antennas requires new, fast and reliable antenna testing methods. In this framework, near-field measurements represent the most interesting solution due to the accuracy and small dimension of the test set compared to far-field and compact range antenna measurement systems.

In near-field measurements [2] the field radiated by the Antenna Under Test (AUT) is measured on a scanning surface placed at short distance ($5\lambda - 7\lambda$, λ being the free space wavelength) from the AUT in an anechoic chamber in order to avoid reflections and stray signals [3]. The propagation process from near-field to far-field conditions is simulated by a proper software. Even if the scanning area could be any sufficiently smooth surface, planar surfaces are most commonly adopted for array antennas.

The main advantages of near-field set-ups are well known: they allow to perform accurate measurements in a controlled environment. However, a further advantage, that is often underestimated, is the possibility of using sophisticated data processing algorithms to reduce the cost of the measurement process without affecting the accuracy. This possibility will be exploited in this paper with reference to array diagnosis.

On the other hand, near-field measurements suffer from two main drawbacks. The first one is the time required to collect the data on the observation surface using standard near-field set-ups [4]. With reference to this point, sparse recovery techniques have been recently proposed in the framework of antenna diagnosis from planar near-field measurements [5–8] in order to reduce the number of measured data and consequently the measurement time. The method has been successfully tested from data acquired in anechoic chamber [9,10].

A further problem is the cost of large anechoic chambers. Regarding this point, a large effort has been devoted to the reduction of the so-called truncation error, caused by a limited scanning area, in order to use smaller and less expensive anechoic chambers [11–14]. However, a further and more drastic solution is to perform measurements in a non anechoic environment, avoiding the use of expensive anechoic chambers.

The aim of this contribution is to investigate a technique that avoids the use of expensive anechoic chambers by filtering out the interference of undesired electromagnetic sources.

It is worth noting that other techniques for filtering interference signals in antenna measurements have been proposed. A partial list is reported in the references section [15–20]. The strategies followed in literature are based on a complete characterization of the environment in order to subtract the environment response [15], on the equivalent source reconstruction using inverse linear approaches [16], on the use of suitable base representations [19,20]. Large effort has also been devoted to interference filtering from amplitude-only measurements [17,18]. Generally speaking, these methods are 'general purpose', in the sense that they can be used in general near-field measurement systems, and do not take explicitly into account the small number of failures in array testing. In the proposed method this characteristic is explicitly exploited to reduce the set of possible array excitations, allowing to identify the failures and to filter the interference contributions in the same step.

The basic idea is to use differential measurements in order to obtain a sparse representation of the AUT excitations [21]. Such sparseness property of the radiating source is used as a-priori information in order to distinguish the AUT contribution from the contribution of scattering objects in measured data, that are characterized by a low rank field distribution.

It is worth noting that the method proposed in this paper strictly resembles the methods used in image processing. In practice, the interference-affected near field measured on a surface is treated as a (complex-data) image. This allows to use some modern image processing algorithms. In particular, two strategies widely used in image processing are applied in this paper. The first one is the reduction of the amount of information by acquiring only the innovation part of an image, as currently happens in video processing, using a differential measurements. The second technique has been recently proposed in video denoising, where the image is split into a low-rank and high-rank part [22]. In particular, in this paper the interference field is filtered out using sparsity as discriminant adopting a mixed minimum ℓ_1 norm and trace norm minimization algorithm.

2. Rank and Sparsity of the Feld Radiated by an Electric Dipole

Before introducing the filtering technique, it is useful to briefly discuss the general idea at the base of the filtering procedure.

Let us consider a harmonic electromagnetic source consisting in an elementary electric dipole directed along the y direction (Figure 1). The dipole can model an element of an array, or also a scattering point caused by objects in the environment where the measurement system is placed. The field of this dipole is observed on a square surface having dimension $L \times L$ ($L = 20\lambda$) placed on the $z = d$ plane with a uniform planar grid at $0.2\ \lambda$ sampling step. The field on the observation points is sampled and the data are collected on an equispaced grid and the measured values are collected in the matrix $\bar{X}$.

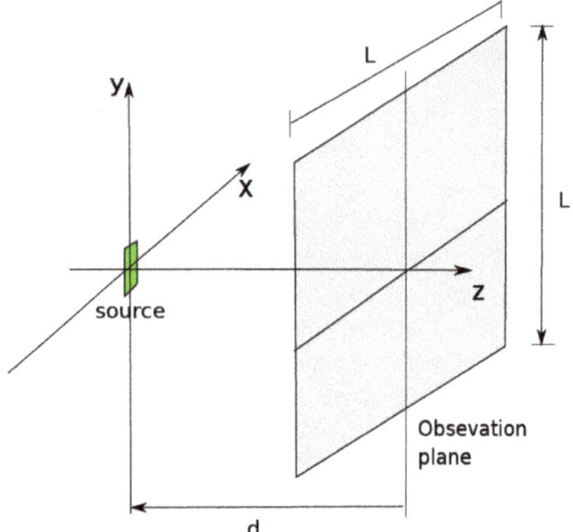

Figure 1. Geometry of the problem.

In Figure 2, the rank of the matrix $\bar{\mathbf{X}}$ is evaluated versus the distance d between the source and the observation plane (blue curve, left scale). We can note that the rank rapidly decreases. As a consequence, at sufficiently large distance $\bar{\mathbf{X}}$ tends to be a low rank matrix.

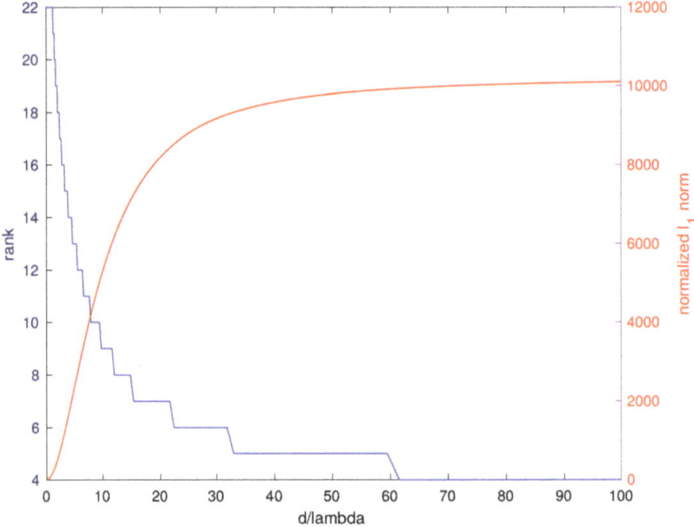

Figure 2. Blue curve (left scale): rank of the field on the observation plane; red curve (right scale): ℓ_1 norm normalized to the maximum of the field amplitude on the observation plane; the observation plane is $20\lambda \times 20\lambda$; d is the distance between the source point and the observation plane.

This 'smoothing' process in the propagation process is a general property of the field radiated by an electromagnetic source [23]. Loosely speaking, the propagation acts as a spatial low pass filter, smoothing the fast spatial variations of the field. Consequently, the field on the observation surface, being smoother, tends to require less basis functions for its representation.

The spreading of the field on the observation surface caused by the filtering property of the propagation phenomenon can also be quantified in terms of 'sparseness' of the field. For this purpose we should evaluate the so called ℓ_0 norm, i.e., the number of elements of the matrix different from zero.

The ℓ_0 norm suffers from a number of drawbacks that prevent its use in practical problems. Instead of the ℓ_0 norm, we will estimate the degree of sparsity using the 1-norm, or ℓ_1 norm, of the matrix $\bar{\mathbf{X}}$, i.e., the sum of all the amplitudes of the entries of the matrix. The more the field is concentrated on the observation surface, the smaller the 1-norm is. This point will be discussed in more detail in the next section. In this section the goal is to give an intuitive explanation of the usefulness of the ℓ_1 norm.

The ℓ_1 norm of the matrix, normalized to the maximum amplitude of the entries of the matrix, is plotted in Figure 2 versus the distance d between the source and the measurement plane (red curve, right scale). We can see that the normalized ℓ_1 norm of the matrix increases rapidly with distance, as we should expect.

This simple example shows that rank and degree of sparsity can be used to distinguish contributions from sources in different positions provided that the plane is positioned in the right position. In particular, if the plane is placed very close to the 'desired' source, and sufficiently far from the interference source, it is possible to filter the undesired sources by subtracting the low rank contribution.

This observation is at the basis of the method proposed in this contribution to filter undesired field reflections.

3. The Array Failure Detection Algorithm with Reflection Filtering

Let us consider an Antenna Under Test (AUT) consisting of a planar $N \times N$ array affected by a number of fault elements (Figure 3). Let Σ be the plane where the AUT aperture lies, and $\mathbf{X}^{AUT} \in \mathbb{C}^{N \times N}$ the matrix collecting the currents of the radiating elements of the AUT.

In differential measurements [5], we consider also an array without failures, called 'golden array', whose currents matrix is $\mathbf{X}^{GOLD}$. $\mathbf{X}^{GOLD}$ (as well as the field radiated by $\mathbf{X}^{GOLD}$) can be obtained by full-wave numerical simulations, or by measurements in a controlled environment (i.e., in an anechoic chamber).

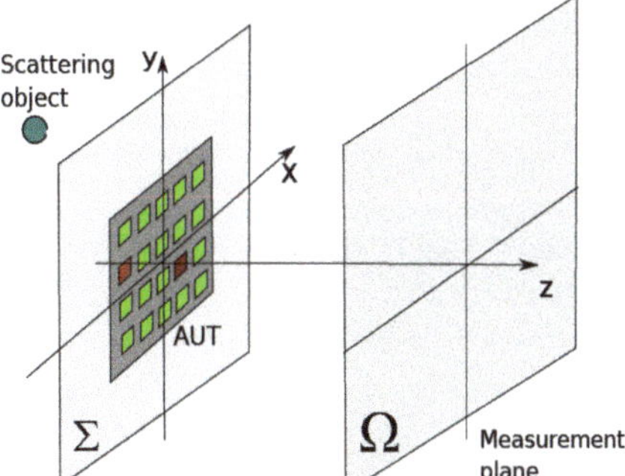

Figure 3. Measurement set-up; the data are collected on the surface Ω placed at a distance d from the AUT and are affected by a scattered field and Gaussian noise; some elements of the AUT are malfunctioning (red squares).

The field of the AUT and of the Golden array are measured on a plane Ω placed at distance d from the AUT in a square lattice of $M \times M$ points. The data are collected in the matrices $\mathbf{Y}^{AUT}$ and $\mathbf{Y}^{GOLD}$, and the following quantities

$$\mathbf{X} = \mathbf{X}^{AUT} - \mathbf{X}^{GOLD} \tag{1}$$

$$\mathbf{Y} = \mathbf{Y}^{AUT} - \mathbf{Y}^{GOLD} \tag{2}$$

are evaluated. Since the number of fault elements is much smaller than the number of elements of the array, the $\mathbf{X}$ matrix is *sparse* [24].

Now, let us suppose that there is an interference source and let us call $\mathbf{X}^s$ the matrix collecting the equivalent currents generated by the field of the interference source on the plane Σ, i.e., the plane where the AUT is placed.

Consequently, the equivalent currents on Σ are given by the superposition of the equivalent currents associated to the AUT and to the interference source, i.e., $\mathbf{X}^{AUT} + \mathbf{X}^s$. Note that on this plane $\mathbf{X}^s$ tends to be a low rank matrix as discussed in the previous section.

The field measured on the observation surface is the superposition of the field radiated by these two contributions plus noise:

$$\mathbf{Y}^m = \mathcal{A}(\mathbf{X}^{AUT} + \mathbf{X}^s) + \mathbf{Y}^n \tag{3}$$

wherein $\mathcal{A}$ is the radiation operator, i.e., the operator mapping the equivalent current matrix on Σ into the matrix collecting the field on Ω [21] and $\mathbf{Y}^n$ is the matrix collecting the measurement noise contribution at the receiver.

The differential measured matrix is consequently:

$$\hat{\mathbf{Y}} = \mathbf{Y}^m - \mathbf{Y}^{GOLD} = \mathcal{A}(\mathbf{X} + \mathbf{X}^s) + \mathbf{Y}^n \tag{4}$$

Rigorously, in order to distinguish the sparse contribution of the AUT and the low rank contribution of the interference source on Σ, the following problem must be solved:

$$\begin{aligned}
&\min \ \ \text{rank}(\mathbf{X}^s) + \alpha\|\mathbf{X}\|_0 \\
&\text{subject to } \|\mathcal{A}(\mathbf{X} + \mathbf{X}^s) - \hat{\mathbf{Y}}\|_2 \le \epsilon
\end{aligned} \tag{5}$$

wherein rank$(\mathbf{X})$ is the rank of the matrix X, $\|\mathbf{X}^s\|_0$ is the ℓ_0 norm of the matrix $\mathbf{X}^s$, α is a regularization parameter and ϵ depends on the level of the noise affecting the data.

Since both rank minimization and ℓ_0 minimization are non convex functions the solution of (5) requires a computational expensive exhaustive search. Furthermore, ℓ_0 norm is instable in presence of noise.

In order to solve the problem it is advantageous to substitute the original problem with a suitably relaxed version.

In particular, the ℓ_0 norm can be substituted by the ℓ_1 entrywise matrix norm [24],

$$\|\mathbf{X}\|_1 = \sum_{k,h} |x_{k,h}| \tag{6}$$

wherein $x_{k,h}$ is the (k, h) entry of the matrix $\mathbf{X}$, while the rank function can be well approximated by the trace norm (also called Schatten 1-norm or nuclear norm) [25,26]:

$$\|\mathbf{X}^s\|_* = \sum_{k=1}^{r} \sigma_k \tag{7}$$

where σ_k is the k-th singular value of $\mathbf{X}^s$ and r is its rank [25,27].

It is interesting to note that nuclear norm and ℓ_1 norm have some similarities since in some way the nuclear norm is to the rank functional what the convex ℓ_1-norm is to the ℓ_0-norm in the sparse recovery area, Figure 4b. In fact, the nuclear norm can be seen as a relaxed version of the rank norm, while the ℓ_1 norm can be considered a relaxed version of the ℓ_0 norm. While ℓ_0 norm counts the number of elements different from zero, the ℓ_1 norm sums up their amplitude. In the same way, while the rank function counts the number of non-zero singular values, the nuclear norm sums their amplitude. In order to clarify this point, let us recall that in sparse recovery the goal is to identify the sparsest vector (i.e., the vector having the largest number of null components) compatible with the available data [24]. This requires to minimize the so-called ℓ_0 norm, wherein ℓ_0 is the number of non null elements of the unknown vector. Such a minimization is a challenging non convex problem. For the sake of simplicity, let us consider a 3 entries vector, $\mathbf{x} = \{x, y, z\}$. The vector is supposed to be 1-sparse, i.e., only one of the three entries of the vector is different from zero. Let us consider the convex hull of the 1-sparse vectors. Such a convex hull turns out to be the unit ball of the ℓ_1 norm, wherein the ℓ_1 norm is $\|\mathbf{x}\|_1 = |x| + |y| + |z|$. A graphical picture of the unit ℓ_1 ball is drawn in Figure 4a. The solution of the ℓ_1 minimization (red point in Figure 4a) is the tangent point between the affine space associated to the available data (drawn as a red line in Figure 4a) and the scaled convex hull. The minimization of the trace norm works in the same way, but operating on the singular values of the matrix.

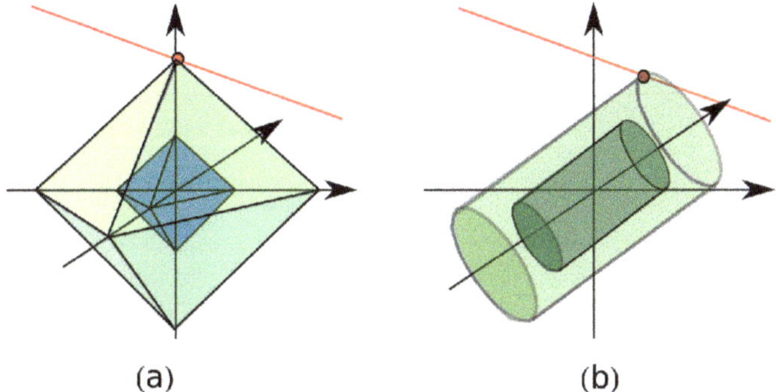

(a)   (b)

Figure 4. (**a**) geometrical picture of the ℓ_1 minimization; (**b**) geometrical picture of the trace norm minimization.

Consequently, in practice the solution of the problem requires the following minimization procedure involving two different definitions of matrix 1-norm:

$$\min \quad \alpha\|\mathbf{X}^s\|_* + \|\mathbf{X}\|_1$$
$$\text{subject to } \|\mathcal{A}(\mathbf{X} + \mathbf{X}^s) - \hat{\mathbf{Y}}\|_2 \leq \epsilon \tag{8}$$

i.e., a weighted minimization of the Schatten 1-norm (i.e., the trace norm) of $\mathbf{X}^s$ and of the entrywise 1-norm of $\mathbf{X}$. The regularization parameter α can be estimated using the L-shape curve adopted also in Tikhonov regularization [28].

The above minimization is a convex problem and can be solved by means of the powerful and efficient algorithms available in many numerical libraries.

The algorithm can handle also multiple scattering interference sources. In this case, the field of each interference source gives a low rank matrix on the observation plane. Consequently, the problem is to identify a set of low-rank matrices and a sparse matrix. i.e.,

$$\min \quad \sum_{l=1}^{L} \alpha_l \|\mathbf{X}_l^s\|_* + \|\mathbf{X}\|_1$$
$$\text{subject to } \left\|\mathcal{A}\left(\mathbf{X} + \sum_{l=1}^{L} \mathbf{X}_l^s\right) - \hat{\mathbf{Y}}\right\|_2 \leq \epsilon \tag{9}$$

where in L is the number of interference sources and $\mathbf{X}_l^s$ is the low rank matrix associated to the l-th interference source.

4. Numerical Examples

In this section some numerical results are shown. The AUT is a 7×7 planar array with $\lambda/2$ inter-element distance, centered on the x, y plane of a Cartesian coordinate system (see Figure 3).

The data are collected on a 21×21 points $\lambda/2$ uniform grid placed on the plane Ω at distance $d = 7\lambda$ from the AUT aperture plane. An undesired source is placed at $\{x = 0, y = 2.2\lambda, z = -8\lambda\}$. The data are affected by -45 dB level Gaussian noise. We suppose that the AUT is affected by three fault elements. The excitation of the non fault elements is one, while the three fault elements have zero excitation. The amplitude of the excitations of the 49 radiating elements of the AUT are plotted in Figure 5a (left figure) in false colors (1 = red, 0 = yellow).

The proposed filtering technique is applied to the measured data. The amplitude of the array excitations is shown in Figure 5c (right figure). The three defects are clearly visible.

As a comparison, the same data have been elaborated using the method [5] consisting of ℓ_1 minimization without filtering procedure. The result is plotted in Figure 5b (central figure), showing a less effective identification of the failures.

In particular, the presence of the undesired source makes two broken elements barely identifiable, while the proposed technique is able to clearly identify all three elements.

The estimation algorithm is also stable compared to the noise level. For example, in Figure 6 the estimation of the failures of the AUT is shown in case of -35 dB noise level. The figure shows that the proposed method still gives acceptable results (Figure 6c, right figure), while the standard method fails to identify at least one failure (Figure 6b, central figure).

In order to show the performance of the algorithm in case of multiple interference sources, two sources placed at $(x = 0, y = 3.2\lambda, z = -8\lambda)$ and $(x = 4, y = 0\lambda, z = -10\lambda)$ are considered. The data are corrupted by -45 dB level Gaussian noise The solution using the filtering technique is shown in Figure 7c, while the solution not implementing the filtering method is shown in Figure 7b, confirming again an improvement in the estimation of the differential excitation.

Finally, an example of identification of the failures in a larger array (81 radiating elements) is reported in Figure 8. The plot shows that standard technique completely fails to identify the fault elements, while the proposed technique is able to identify the area where two fault elements are placed. The position of the third fault element is not detected, but the figure shows a variation of the excitations on the left upper corner of the array.

In Table 1 the results of the simulations are briefly compared in a quantitative way. As figure of merit, the Mean Square Error between the reference differential amplitude excitations of the AUT and the amplitude excitation of the retrieved differential excitations are reported using the interference source filtering algorithm (4th column and using standard algorithm (5th column of the Table). The CPU time required by the filtering algorithm is reported in the 6th column. Loosely speaking, also the quantitative parameter chosen shows an improvement in the differential excitation reconstruction using the proposed algorithm. The improvement becomes more relevant increasing the number of interference sources. For example, in the case of two sources, the MSE decreases from 1.5 dB to -3.7 dB. Even if this improvement is numerically lower than the 1 source case, it is practically much more relevant, and allows to pass from no failure detection to an effective failure detection, as shown in the previous section. Regarding the computation time, the examples were obtained on an Mc Air 11' with i7 processor using CVX using only one core. The computation time is less than two minutes, and increases almost linearly with the number of interference sources. However, these values are only indicative, and can give an erroneous idea of the computational time required in real applications.

Table 1. First column: number of the example; second column: number of the elements of the AUT; third column: number of interference sources; fourth column: Mean Square Error of the amplitude of the differential excitations using the proposed technique; fifth column: Mean Square Error of the amplitude of the differential excitations without using the proposed technique; sixth column: CPU time (seconds) required by the filtering program.

Example	Array Elements	Number of Interf.	MSE Filt	MSE no Filt.	CPU Time
1st	7×7	1	-7.5 dB	-3.7 dB	95 s
2rd	7×7	1	-9.5 dB	-3.7 dB	77 s
3rd	7×7	2	-3.7 dB	1.5 dB	167 s
4rth	9×9	1	-4.8 dB	3.9 dB	78 s

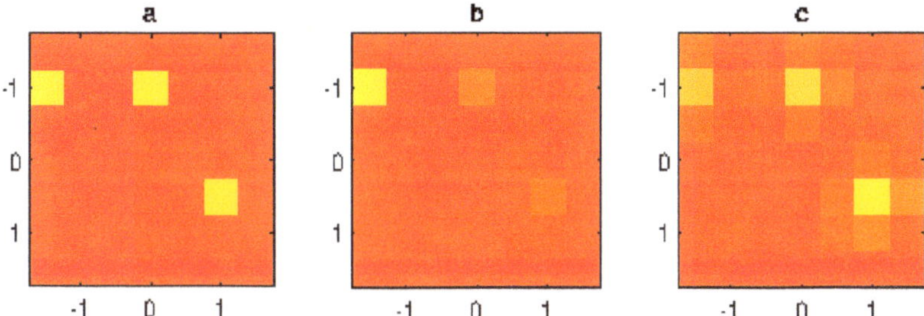

Figure 5. 1st example: normalized excitation amplitude of the radiating elements (linear scale in false colors: yellow = null amplitude, red = unit amplitude); (**a**) exact array excitations; (**b**) excitations obtained without filtering; (**c**) excitations obtained using the proposed filtering method; 7×7 planar array with $\lambda/2$ inter-element distance, 21×21 measurement points, $d = 7\lambda$, measured data affected by interference field radiated by a source placed at $(x = 0, y = 2.2\lambda, z = -8\lambda)$ and by -45 dB level Gaussian noise.

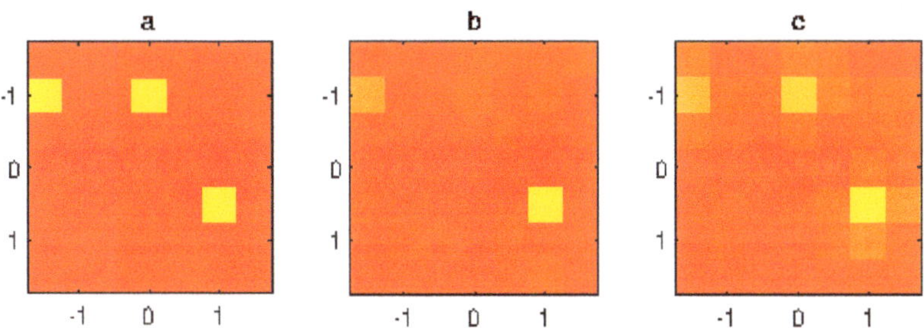

Figure 6. 2nd example;: normalized excitation amplitude of the radiating elements (linear scale in false colors: yellow = null amplitude, red = unit amplitude); (**a**) exact array excitations; (**b**) excitations obtained without filtering; (**c**) excitations obtained using the proposed filtering method; 7×7 planar array with $\lambda/2$ inter-element distance, 21×21 measurement points, $d = 7\lambda$, measured data affected by interference field radiated by a source placed at $(x = 0, y = 2.2\lambda, z = -8\lambda)$ and by -35 dB level Gaussian noise.

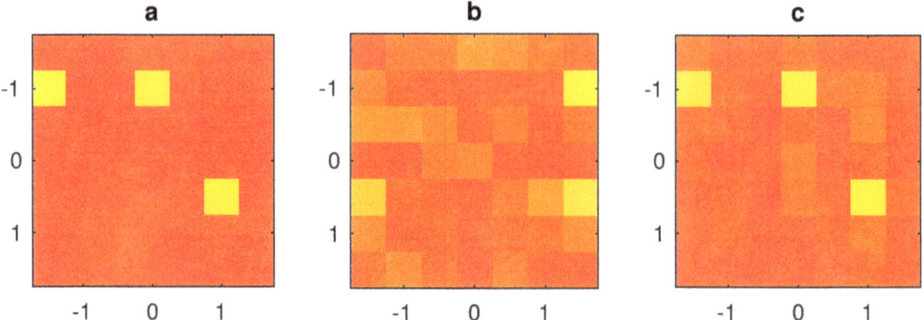

Figure 7. 3rd example: normalized excitation amplitude of the radiating elements (linear scale in false colors: yellow = null amplitude, red = unit amplitude); (**a**) exact array excitations; (**b**) excitations obtained without filtering; (**c**) excitations obtained using the proposed filtering method; 7×7 planar array with $\lambda/2$ inter-element distance, 21×21 measurement points, $d = 7\lambda$, measured data affected by interference field radiated by a source placed at $(x = 0, y = 3.2\lambda, z = -8\lambda)$ and a source placed at $(x = 4, y = 0\lambda, z = -10\lambda)$. The data are corrupted by -45 dB level Gaussian noise.

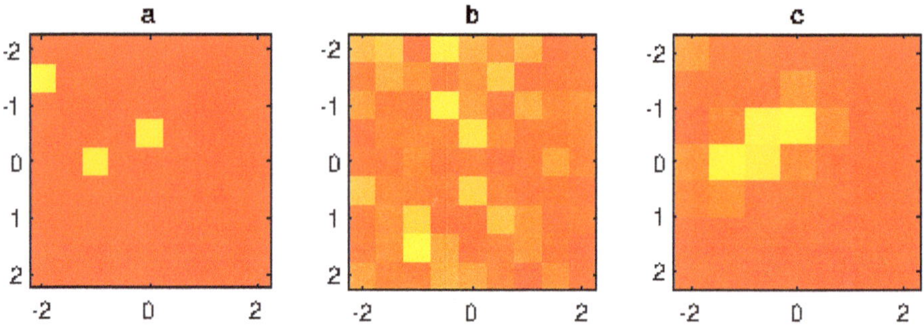

Figure 8. 4th example: normalized excitation amplitude of the radiating elements (linear scale in false colors: yellow = null amplitude, red = unit amplitude); (**a**) exact array excitations; (**b**) excitations obtained without filtering; (**c**) excitations obtained using the proposed filtering method; 9×9 planar array with $\lambda/2$ inter-element distance, 21×21 measurement points, $d = 7\lambda$, measured data affected by interference field radiated by a source placed at $(x = 0, y = 3.2\lambda, z = -8\lambda)$ and by -45 dB level Gaussian noise.

5. Conclusions

In this paper a novel filtering algorithm of signals in planar near-field measurements is described. The method allows the filtering strategy of interference sources in array diagnosis. The technique is simple and numerically efficient, since it allows the use of convex minimization procedures.

The basic idea is to take advantage of the characteristics of the electromagnetic propagation in terms of 'spreading' of the field distribution. Briefly, the equivalent current distribution on the array is strongly concentrated on the radiating elements, while the contribution of interference sources placed far from the plane of the array is smoother. Accordingly, it is possible to distinguish these two contributions looking for a 'sparse' distribution and a 'low rank' distribution. Numerical examples carried out in some simple cases confirm the effectiveness of this approach.

As discussed in the paper, the method proposed in this paper strictly resembles the methods used in image processing. In practice, the interference-affected near field measured on a surface is treated as a (complex-data) image. This allows to use some modern image processing algorithms. In particular, two strategies widely used in image processing are applied in this paper. The first one is the reduction the amount of information by acquiring only the innovation part of an image, as currently happens in video processing, using differential measurements. The second technique has been recently proposed in video denoising and splits the image into a high-sparse and high-rank part.

The results have been obtained using a small laptop computer and CVX program. As stressed in the Introduction, the aim of this paper is to introduce the technique, and for this purpose the computer and program adopted are acceptable. However, CVX is a slow program developed mainly for research-stage applications. More powerful and efficient algorithms are available under payment. The use of these algorithms on powerful parallel computers drops the computational time drastically. Even if computational time is not an issue in near-field measurements, since it is usually a fraction compared to the time required for data acquisition, the possibility of fast reconstruction opens the thrilling possibility of online failure identification, i.e., identification of failures on-site while antenna works [29], filtering the environmental noise of the site where the antenna is placed. This interesting possibility encourages to continue the investigation on the rank properties of the field in the framework of antenna measurements.

Author Contributions: Conceptualization M.D.M.; software, D.P. and M.L.; formal analysis G.P. and F.S.

Funding: This work was supported by the MIUR Program Dipartimenti di Eccellenza 2018–2022.

References

1. Agiwal, M.; Roy, A.; Saxena, N. Next Generation 5G Wireless Networks: A Comprehensive Survey. *IEEE Commun. Surv. Tutor.* **2016**, *18*, 1617–1655. [CrossRef]
2. Yaghjian, A. An overview of near-field antenna measurements. *IEEE Trans. Antennas Propag.* **1986**, *34*, 30–45. [CrossRef]
3. Newell, A.C. Error analysis Techniques for Planar Near-Field Measurements. *IEEE Trans. Antennas Propag.* **1988**, *36*, 754–768. [CrossRef]
4. Migliore, M.D. Near Field Antenna Measurement Sampling Strategies: From Linear to Nonlinear Interpolation. *Electronics* **2018**, *7*, 257. [CrossRef]
5. Migliore, M.D. A Compressed Sensing Approach for Array Diagnosis From a Small Set of Near-Field Measurements. *IEEE Trans. Antennas Propag.* **2011**, *59*, 2127–2133. [CrossRef]
6. Oliveri, G.; Carlin, M.; Massa A. Complex-Weight Sparse Linear Array Synthesis by Bayesian Compressive Sampling. *IEEE Trans. Antennas Propag.* **2012**, *60*, 2309–2326. [CrossRef]
7. Oliveri, G.; Rocca, P.; Massa, A. Reliable diagnosis of large linear arrays—A bayesian compressive sensing approach. *IEEE Trans. Antennas Propag.* **2012**, *60*, 4627–4636. [CrossRef]
8. Salucci, M.; Gelmini, A.; Oliveri, G.; Massa, A. Planar arrays diagnosis by means of an advanced Bayesian compressive processing. *IEEE Trans. Antennas Propag.* **2018**, *66*, 5892–5906. [CrossRef]
9. Fuchs, B.; Le Coq, L.; Migliore, M.D. Fast Antenna Array Diagnosis from a Small Number of Far-Field Measurements. *IEEE Trans. Antennas Propag.* **2016**, *64*, 2227–2235. [CrossRef]
10. Costanzo, S.; Borgia, A.; di Massa, G.; Pinchera, D.; Migliore, M.D. Radar Array Diagnosis from Undersampled Data Using a Compressed Sensing/Sparse Recovery Technique. *J. Electr. Comput. Eng.* **2013**, *2013*, 627410. [CrossRef]
11. Bolomey, J.; Bucci, O.M.; Casavola, L.; D'Elia, G.; Migliore, M.D.; Ziyyat, A. Reduction of truncation error in near-field measurements of antennas of base-station mobile communication systems. *IEEE Trans. Antennas Propag.* **2004**, *52*, 593–602. [CrossRef]
12. Salucci, M.; Migliore, M.D.; Oliveri, G.; Massa, A. Antenna measurements-by-design for antenna qualification. *IEEE Trans. Antennas Propag.* **2018**, *66*, 6300–6312. [CrossRef]
13. Martini, E.; Breinbjerg, O.; Maci, S. Reduction of Truncation Errors in Planar Near Field Aperture Antenna Measurements Using the Gerchberg-Papoulis Algorithm. *IEEE Trans. Antennas Propag.* **2008**, *56*, 3485–3493. [CrossRef]

14. Migliore, M.D.; Salucci, M.; Rocca, P.; Massa, A. Truncation-Error Reduction in Antenna Near-Field Measurements Using an Overcomplete Basis Representation. *IEEE Antennas Wirel. Propag. Lett.* **2019**, *18*, 283–287. [CrossRef]

15. Burnside, W.D.; Gupta, I.J. A method to reduce stray signal errors in antenna pattern measurements. *IEEE Trans. Antennas Propag.* **1994**, *42*, 399–405. [CrossRef]

16. Alvarez, Y.; Las-Heras, F.; Pino, M.R. The sources reconstruction method for amplitude-only field measurements. *IEEE Trans. Antennas Propag.* **2010**, *58*, 2776–2781. [CrossRef]

17. Brown, T.; Jeffrey, I.; Mojabi, P. Multiplicatively regularized source reconstruction method for phaseless planar near-field antenna measurements. *IEEE Trans. Antennas Propag.* **2017**, *65*, 2020–2031. [CrossRef]

18. Quijano, J.L.A.; Vecchi, G. Field and source equivalence in source reconstruction on 3D surfaces. *Prog. Electromagn. Res.* **2010**, *103*, 67–100. [CrossRef]

19. Bucci, O.M.; D'Elia, G.; Migliore, M.D. A general and effective clutter filtering strategy in near-field antenna measurements. *Proc. IEE Microw. Antennas Propag.* **2004**, *151*, 227–235. [CrossRef]

20. Quijano, J.L.A.; Vecchi, G.; Li, L.; Sabbadini, M.; Scialacqua, L.; Bencivenga, B.; Mioc, F.; Foged, L.J. 3D spatial filtering applications in spherical near field antenna measurements. In Proceedings of the AMTA 2010 Symposium, Atlanta, GA, USA, 10–15 October 2010.

21. Migliore, M.D. On the Sampling of the Electromagnetic Field Radiated by Sparse Sources. *IEEE Trans. Antennas Propag.* **2015**, *63*, 553–564. [CrossRef]

22. Ji, H.; Liu, C.; Shen, Z.; Xu, Y. Robust video denoising using low rank matrix completion. In Proceedings of the IEEE Computer Society Conference on Computer Vision and Pattern Recognition, San Francisco, CA, USA, 13–18 June 2010; pp. 1791–1798.

23. Migliore, M.D. Minimum Trace Norm Regularization (MTNR) in Electromagnetic Inverse Problems. *IEEE Trans. Antennas Propag.* **2016**, *64*, 630–639. [CrossRef]

24. Migliore, M.D. A simple introduction to compressed sensing/sparse recovery with applications in antenna measurements. *IEEE Antennas Propag. Mag.* **2014**, *56*, 14–26. [CrossRef]

25. Candes, E.J.; Tao, T. The Power of Convex Relaxation: Near-Optimal Matrix Completion. *IEEE Trans. Inf. Theory* **2010**, *56*, 2053–2080. [CrossRef]

26. Fuchs, B.; Le Coq, L.; Migliore, M.D. On the Interpolation of Electromagnetic Near Field without Prior Knowledge of the Radiating Source. *IEEE Trans. Inf. Theory* **2017**, *65*, 3568–3574. [CrossRef]

27. Keshavan, R.H.; Montanari, A.; Oh, S. Matrix Completion From a Few Entries. *IEEE Trans. Inf. Theory* **2010**, *56*, 2980–2998. [CrossRef]

28. Brancaccio, A.; Migliore, M.D. A Simple and Effective Inverse Source Reconstruction with Minimum a Priori Information on the Source. *IEEE Geosci. Remote Sens. Lett.* **2017**, *14*, 454–458. [CrossRef]

29. Migliore, M.D.; Pinchera, D.; Lucido, M.; Schettino, F.; Panariello, G. A Sparse Recovery Approach for Pattern Correction of Active Arrays in Presence of Element Failures. *IEEE Antennas Wirel. Propag.* **2015**, *14*, 1027–1030. [CrossRef]

 Journal of
Imaging

Article

Qualitative Methods for the Inverse Obstacle Problem: A Comparison on Experimental Data

Martina T. Bevacqua [1,2,*,†] and **Roberta Palmeri** [1,†]

[1] DIIES, Department of Information Engineering, Infrastructures and Sustainable Energy, Università Mediterranea di Reggio Calabria, via Graziella, Loc. Feo di Vito, 89124 Reggio Calabria, Italy; roberta.palmeri@unirc.it

[2] CNR-IREA, National Research Council of Italy, Institute of Electromagnetic Sensing of the Environment, via Diocleziano 328, 80124 Napoli, Italy

[*] Correspondence: martina.bevacqua@unirc.it; Tel.: +39-09-651-693-262

[†] The paper is a result of a joint contribution of both the authors, from conceptualization to final writing and editing.

Received: 7 March 2019; Accepted: 8 April 2019; Published: 10 April 2019

Abstract: Qualitative methods are widely used for the solution of inverse obstacle problems. They allow one to retrieve the morphological properties of the unknown targets from the scattered field by avoiding dealing with the problem in its full non-linearity and considering a simplified mathematical model with a lower computational burden. Very many qualitative approaches have been proposed in the literature. In this paper, a comparison is performed in terms of performance amongst three different qualitative methods, i.e., the linear sampling method, the orthogonality sampling method, and a recently introduced method based on joint sparsity and equivalence principles. In particular, the analysis is focused on the inversion of experimental data and considers a wide range of (distinct) working frequencies and different kinds of scattering experiments.

Keywords: inverse obstacles problem; inverse source problem; joint sparsity; linear sampling method; microwave imaging; orthogonality sampling method

1. Introduction

The inverse scattering problem aims to retrieve the electromagnetic properties of unknown targets by processing the field they scatter [1,2]. This problem can be of interest in biomedical imaging in order to discriminate between healthy and malignant biological tissues, identify anomalies inside the human body, and build an electromagnetic model of a specific anatomical district [3–5]. Moreover, it is relevant in other applications, such as underground prospecting as well as non-destructive testing of materials, in order to detect and characterize buried targets and internal defects [6,7].

Due to the non-linearity and ill-posedness of this class of inverse problems [1,8], the quest to develop reliable, accurate, and effective solution methods is still ongoing. As witnessed by many papers published in literature, several researchers and scholars are very active on this topic, covering its challenging computational, algorithmic, modeling, and experimental aspects.

Due to the mathematical difficulties of the problem, many studies have been focused on the introduction of solution methods for the corresponding inverse obstacle problem, whose scope is to recover just the morphological information about the targets, namely their supports, by giving up their electromagnetic properties [9–16]. Such methods are known as qualitative methods and are different from quantitative methods, which aim at retrieving both electrical and morphological properties of the region of interest [2,9]. Indeed, they tackle the problem in its full non-linearity, without involving any approximation and allowing for a widening of the class of retrievable objects; unfortunately, such methods are characterized by a higher computational burden and longer processing time [2]. On the other hand, qualitative methods are characterized by simplified mathematical models and a lower computational burden [9].

Among qualitative methods, it is worth mentioning sampling methods that include the linear sampling method (LSM) [12] (and the related factorization method [13]), as well as the orthogonality sampling method (OSM) [14]. The main idea of these methods is that of sampling the investigation domain in an arbitrary grid of points and computing at each point an indicator function whose energy will assume different values depending on whether the sampled point belongs or not to the scatter. Although both LSM and OSM belong to the class of sampling methods, they exhibit different features. For instance, the LSM indicator function is computed by solving an auxiliary linear problem whose solution involves the adoption of a regularization technique. In a different fashion, the OSM does not require the solution of a linear problem since the indicator function is just related to the evaluation of a scalar product. Moreover, the OSM seems to be more flexible with respect to the measurement configurations.

By taking advantage of recent results in the area of compressive sensing and sparsity promoting techniques [17], a new qualitative method has been very recently introduced in the literature that, provided some conditions hold true, is able to retrieve the boundary of unknown targets rather than their support [15,16]. Such a method exploits the equivalence theorem in the case of dielectric obstacles and takes advantage of the particular distribution assumed by the induced currents in case of perfect electric conductors [18]. Differently from sampling methods, it does not sample the investigation domain in a grid of points, and it does not require the selection of a fixed threshold to discriminate between points inside and outside the targets. However, if compared with LSM and OSM, it shows a heavier computational burden.

Encouraged by the remarkable properties of the methods discussed above, in the following, some comparisons and performance analysis are provided. In particular, the tests are performed against experimental data [19] and consider a wide range of working frequencies and different kinds of (monochromatic) data.

The paper is organized as follows. In Section 2, the inverse scattering problem is formulated. Moreover, a brief summary of the three considered qualitative methods is given, including a discussion on their expected advantages and disadvantages. In Section 3, some examples considering experimental Fresnel data are reported. Finally, conclusions follow. Throughout the paper we consider the canonical 2D scalar problem (transverse magnetic (TM) polarized fields) and we assume and drop the time harmonic factor $exp\{j\omega t\}$.

2. Materials and Methods

2.1. Mathematical Formulation of the Problem

Let D denote the region being tested where targets with support Σ are located. Let $\chi(r) = \epsilon_s(r)/\epsilon_b - 1$ be the contrast function that relates the electromagnetic properties of the scatterers to those of the host medium, wherein ϵ_s and ϵ_b are the complex permittivities of the scatterers and the background medium, respectively. Let $E_s^\infty(\hat{r}_m, \hat{r}_t)$ be the scattered far-field pattern measured on a closed circle curve Γ in the far zone of the scatterers in direction $\hat{r}_m$ when a unit amplitude plane wave impinges from direction $\hat{r}_t$. Then, the equations describing the scattering problem can be expressed as [1]:

$$E_s^\infty(\hat{r}_m, \hat{r}_t) = \int_D G_b^\infty(\hat{r}_m, r)W(r', \hat{r}_t)dr' = \mathcal{A}_e[W] \tag{1}$$

$$E(r, \hat{r}_t) = E_i(r, \hat{r}_t) + \int_D G_b(r', r)W(r', \hat{r}_t)dr' = E_i + \mathcal{A}_i[W] \tag{2}$$

where E_i and E are the incident and total field, respectively, $W = \chi E$ is the contrast source, and $r = (x, y)$ scans the investigation domain D. Moreover, G_b^∞ is the Green's function in the far-field zone pertaining to the background medium. Finally, $\mathcal{A}_e$ and $\mathcal{A}_i$ are short notation for the integral radiation operators.

In the inverse obstacle problem, one aims at retrieving the support Σ of the target, by giving up its electromagnetic properties, from the (noise corrupted) measured data E_s^∞ when the targets are illuminated by a known incident field E_i [9]. Such a problem is non-linear as well as ill-posed because of the properties of the involved operator $\mathcal{A}_e$ [1,8].

2.2. Linear Sampling Method

The LSM is one of the most popular qualitative methods to retrieve objects' support from the measurements of the corresponding scattered field data [12]. More in detail, it consists in sampling the region under test into an arbitrary grid of points and solving for each point the so-called *far field integral equation* given by:

$$\int_\Gamma E_s^\infty(\hat{r}_m, \hat{r}_t)\, \xi(r_s, \hat{r}_t)\, d\hat{r}_t = G_b^\infty(\hat{r}_m,\, r_s) \tag{3}$$

wherein ξ represents the unknown function and r_s the sampling point. Despite the linearity of Equation (3), the evaluation of the solution ξ is not straightforward due to the compactness of the operator at the left hand side, which implies ill-posedness in its inversion [12,20]. Hence, a regularization technique is required. An effective choice is that of solving it by adopting the Tikhonov regularization [20]. Then, the estimation of the unknown support is pursued by evaluating the l_2 norm (i.e., the "energy") of the unknown function ξ with respect to r_t, given by:

$$I_{LSM}(r_s) = \|\xi(r_s, \hat{r}_t)\|_2^2 = \int_\Gamma \left|\xi(r_s, \hat{r}_t)\right|^2 d\hat{r}_t \tag{4}$$

The above defined indicator depends on the sampling point r_s and assumes a finite value when the sampling point belongs to the unknown object, while it blows up elsewhere [12]. As such, by selecting a fixed threshold, the indicator described in Equation (4) allows to discriminate between points inside and outside the scatterers and finally retrieve Σ [20]. It is worth noting that the computational burden is low, as the solution of Equation (4) just involves the evaluation of the singular values decomposition (SVD) of the measured data [20].

2.3. Orthogonality Sampling Method

The OSM has recently been introduced in the literature [14]. It consists in computing the reduced scattered field from the far-field pattern $E_s^\infty(r_m, r_t)$. Such a reduced field can be computed by evaluating the scalar product between the far field and the Green's function in far-field zone, that is [14]:

$$E_s^{red}(r_s, \hat{r}_t) = \int_\Gamma E_s^\infty(\hat{r}_m, \hat{r}_t)\, e^{jk_b\, r_s\cdot\hat{r}_m}\, d\hat{r}_m = \frac{1}{\gamma}\langle E_s^\infty, G_b^\infty\rangle_\Gamma \tag{5}$$

where $\langle\cdot,\cdot\rangle$ denotes the scalar product and γ is a constant for a fixed frequency [14]. In a nutshell, Equation (5) tests the orthogonality relation between the far-field pattern and the Green's function in the far-field zone (apart from a constant). Then, the OSM indicator $I_{OSM}(r_s)$ is defined as:

$$I_{OSM}(r_s) = E_s^{red}\|r_s, \hat{r}_t\|_2^2 = \int_\Gamma \left|E_s^{red}(r_s, \hat{r}_t)\right|^2 d\hat{r}_t \tag{6}$$

and achieves large values when the far-field pattern approaches the one for the background Green's function for the considered sampling point [14]. Again, by selecting a fixed threshold, it is possible to discriminate sampling points belonging to the target support [14].

2.4. Shape Reconstruction Via Joint Sparsity Based Inverse Source and Equivalence Principles

In order to deal with inverse obstacle problems, one can also take advantage of the joint solution of a number of related inverse source problems that consist of recovering the induced currents from the knowledge of the measured far field pattern E_s^∞ [15,16]. The strategy is obviously suggested by the fact that the support of W is exactly equal to Σ whatever incident field E_i is. Note that inverse source problems are linear (see Equation (1)) but are still (severely) ill-posed such that some additional care is needed.

In this respect, note that when a generic electromagnetic field is propagating in the space in the presence of dielectric obstacles, it induces some currents inside the obstacles, which are expected to be different from zero inside Σ and in turn become sources of the scattered field data. However, by virtue of the equivalence theorem [18] (which is a key point of the approach), the measured scattered field can be eventually conceived as the one radiated by some equivalent surface currents, i.e., electric W_s and magnetic W_{smx} and W_{smy} surface currents, which can play the role of the original ones.

The main advantage of considering the equivalent currents rather than the original ones is that they are distributed over the boundary of Σ. As a consequence, they can be identified by only a few non-zero pixels in the investigation domain and, hence, can be considered sparse with respect to the pixel basis representation. Then, a possible idea for their retrieval is that of looking for the sparsest current distributions, which contemporarily satisfy the measured data. To this end, sparsity promoting approaches can be exploited [17].

Interestingly, the property of sparsity displayed by the equivalent currents holds true regardless of the direction of illumination $\hat{r}_t$ or the kind of measurement configuration. In order to enforce this joint sparsity among the different scattering experiments, an auxiliary variable $\mathcal{B}$ can be defined as the common upper bound on the amplitude of the equivalent currents for the different scattering illumination conditions. Accordingly, the problem is recast as [15,16]:

$$\min\|\mathcal{B}(r)\|_1 \tag{7a}$$

$$s.t. \quad \left\|E_s^\infty - \mathcal{A}_e[W_s] - \mathcal{A}_{mx}[W_{smx}] - \mathcal{A}_{my}[W_{smy}]\right\|_2 \leq \delta \tag{7b}$$

$$\begin{aligned}\left|W_s(r,\hat{r}_t)\right| &\leq \mathcal{B}(r) \\ \left|W_{smx}(r,\hat{r}_t)\right| &\leq \mathcal{B}(r), \quad \forall\, \hat{r}_t \in \Gamma \\ \left|W_{smy}(r,\hat{r}_t)\right| &\leq \mathcal{B}(r)\end{aligned} \tag{7c}$$

where $\|\cdot\|_1$ is the l_1 norm, and $\mathcal{A}_{mx}$ and $\mathcal{A}_{my}$ are the integral radiation operators that relate the magnetic surface currents to the scattered electric fields.

The variable $\mathcal{B}$ does not depend on the direction of illumination $\hat{r}_t$, but only on the position r in the investigation domain. Moreover, it is expected to assume non-zero values only on the boundary of Σ such that it will act as a "boundary" indicator. For this reason, the method is called *boundary retrieval through inverse source and sparsity* (B-IS). As a consequence, differently from LSM and OSM, there is no need for a thresholding to retrieve the support from the indicator map, as the approach directly retrieves the boundary of the targets.

Note the approach can be also applied to the case of perfect electrical conductors. In fact, in such a case, the induced currents W exist only on the boundary of Σ and there is no need for introducing equivalent surface currents. Then, the optimization problem (7) is further simplified [15,16].

As a final comment, note that the B-IS method belongs to the class of the joint sparsity-promoting procedures. In particular, the joint sparsity used herein is not enforced by means of the mixed norm $l_1 - l_2$ as in References [21,22], but through the definition of the auxiliary variable $\mathcal{B}$.

2.5. Applicability and Limititations of the Three Methods

In LSM, the far field Equation (3) can be interpreted as an attempt to synthetize induced currents such that their radiating part is focused and/or circularly symmetric with respect to the given sampling point [20,23]. As such, it correctly works as long as the object is convex and the non-radiating component of the induced currents is negligible [20]. Usually, when the product of the maximum amplitude of the contrast χ times the maximum size of the target is large, which corresponds to an increasing amount of non-radiating sources, it does not guarantee reliable reconstructions.

The OSM is based on a test of orthogonality, which allows one to compute the reduced scattered field. As far as its physical interpretation is concerned, it has been recently argued in Reference [24] that the reduced scattered field is related to the radiative part of the currents. As such, similarly to LSM, the OSM indicator does not take into account the non-radiating part of the currents, so that, again, limitations are expected with increasing electrical dimensions of the scatterer.

While both the LSM and OSM act on volumetric currents, which are located on all points belonging to the support of the target, B-IS aims at jointly solving several inverse source problems, wherein one is looking for currents that, under some reasonable hypotheses, are located on the boundary of Σ. Notably, the optimization problem (7) does not aim at retrieving exactly the equivalent currents but rather to find a common upper bound to their amplitudes. Also note the possible (superficial) non-radiating currents do not seem to play any role in the process, while the possible non-radiating volumetric currents are filtered out by the sparsity regularization. For all the above, the B-IS method can be understood as more robust with respect to the presence of non-radiating currents than LSM and OSM. Moreover, the upper bound in Equation (7) can be understood as a trade-off between sparsity promoting and data fitting. As a consequence, B-IS will be able to retrieve at least the convex-hull of the targets. Remarkably, probably because of the use of the l_1 norm rather than the l_0 pseudo-norm, one can also be able to retrieve the shape of some non-convex target [15]. As a consequence of all the above, B-IS is expected to fail when the boundary of the target is not sparse (for instance in case of a ragged boundary), or when, according to the sparsity theory [17], the number of measurements is not sufficiently large with respect to the degree of sparsity of the target contour.

Concerning the requirements about the kind of measured data, the LSM needs to correctly discretize the integral operator in Equation (3) by collecting a sufficient number of data in multiview-multistatic measurement configuration. As a consequence, even if the product of the maximum amplitude of the contrast χ times the maximum size of the target is not large, the undersampling of the measured data can compromise the reconstructions. On the other side, the OSM requires an accurate discretization of the integral in Equation (5) over the measurements curve, and hence, a sufficient number of measurements. However, differently from LSM, OSM can compensate the low number of data points by considering other kind of experiments different from multiview-multistatic ones. In fact, OSM and B-IS exhibit wide flexibility with respect to processed data and can offer the possibility of combining information arising from frequency diversity without the need of a posteriori merging the single results. To this end, the OSM indicator can be easily modified by integrating over a given frequency band [14,24], and the B-IS indicator function implicitly defined by Equation (7c) can be redefined as the upper bound for the different sources at the different frequencies [15].

Another important difference among the methods regards the computational burden. The OSM indicator does not involve the solution of an ill-posed problem but only the evaluation of a scalar product. This implies more robustness with respect to measurement errors on the scattered data and a very low computation burden. On the contrary, the LSM indicator function is computed by solving an auxiliary linear problem, whose solution involves the adoption of a regularization technique. However, the main computational effort of the method is the evaluation of the SVD of measured data, whose dimension is only dictated by the number of scattering experiments. This task has to be carried out only once, since the kernel of the linear system is the same for all sampling points. With respect to the LSM and the OSM, B-IS exhibits a heavier computational burden, which drastically increases with the number of unknowns, since an l_1 norm optimization problem has to be solved. On the other side, the problem underlying the B-IS belongs to the class of convex programming problems, whose single optimal solution can be reached using any common local optimization procedure.

In terms of reconstruction accuracy, both the LSM and OSM require the selection of a fixed threshold to discriminate between points inside and outside the targets. In a different fashion, the B-IS does not provide an indicator map but directly looks for the boundary of the target. As such, it is expected to be more accurate.

These aspects, which are briefly summarized in Table 1, will be further investigated in the following numerical section.

Table 1. Summary of strong and weak points of the three solution methods.

	LSM	OSM	B-IS
Reconstruction accuracy	medium	medium	high
Computational burden	low	very low	high
Flexibility with respect to the kind of data	low	high	high

3. Results

In this section, an accurate comparative study was performed against the experimental data set provided by the Institute Fresnel of Marseille [19], typically adopted to benchmark inverse scattering procedures. In particular, four targets from the 2001 Fresnel database were considered, i.e.,:

- the DielTM target, which consists of a dielectric homogeneous cylinder of radius 1.5 cm and relative permittivity 3 ± 0.3
- the RectTM_Dece target, which is a rectangular metallic target of 25.4 mm × 12.7 mm not centered with respect to the azimuthal positioner axis;
- the U-TM shaped target, which is a metallic U-shaped target with dimension 80×50 mm^2
- the TwinDielTM target, which consists of two identical dielectric homogeneous cylinders of radius 1.5 cm and relative permittivity 3 ± 0.3.

The complete description of the targets and the measurement set-up can be found in Reference [19], while a schematic sketch is shown in Figure 1. The data were collected under a partially aspect-limited configuration, where primary sources completely surrounded the targets, but for each illumination, the measurements were taken only on an angular sector of 240° achieved by excluding the 120° sector centered on the incidence direction.

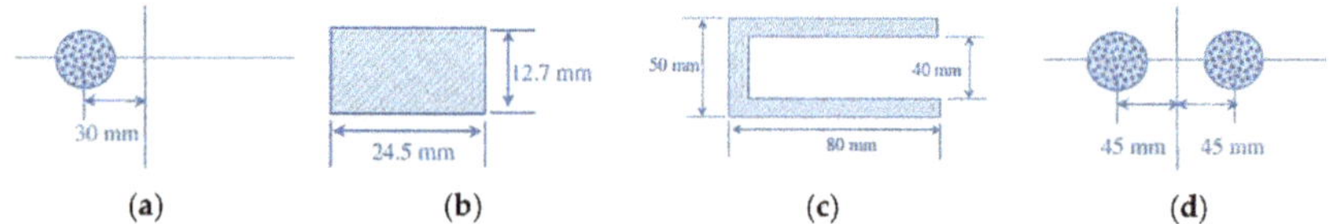

Figure 1. Targets from 2001 Fresnel dataset: (**a**) DielTM target; (**b**) RectTM_Dece target; (**c**) U-TM-shaped target, and (**d**) TwinDielTM target.

For the dielectric targets, the working frequencies could change among the interval 1–8 GHz with a step of 1 GHz. Also, 12 GHz and 16 GHz data were available. On the other side, for the metallic targets the working frequencies can change among the interval 2–16 GHz with a step of 2 GHz. For the sake of brevity, only some significant reconstructions corresponding to some frequencies will be shown in the following. Moreover, for each target, two different datasets have been considered. The first one consisted of a 36 × 36 multiview-multistatic data matrix (the number of measurements and views M is equal to 36), while the second one corresponded to a data matrix with halved dimensions, that is M equal to 18. The investigated area of 0.1775 m × 0.1775 m was discretized in 40 × 40 cells.

In the following examples the single frequency indicators were normalized with respect to their maximum values. In particular, the LSM indicator were rescaled as described in Reference [25].

Note that the number of independent data were lower with respect to the numbers of unknowns. This is a common problem of all inverse scattering and inverse obstacles problems. In fact, one is not able to collect data at will, as only a finite number of independent experiments can be performed [26]. However, the qualitative methods we were comparing did not look for a quantitative reconstruction in every pixel of the scenario, but they looked for some qualitative behavior of the indicator. Moreover, all three methods somehow exploited additional regularizations.

3.1. Convex Dielectric Target

The results corresponding to the single dielectric circular cylinder are reported in Figure 2. As can be seen, at the lowest frequencies, the three methods could localize the cylinder (whose diameter was $0.1\lambda_b$ at 1 GHz and $0.2\lambda_b$ at 2 GHz (λ_b was the wavelength in free space)). Its size was overestimated, especially by LSM and OSM, while instead the B-IS method was not able to correctly identify its shape. In the frequency range of 3–8 GHz, the performance of the three methods improved and the support was correctly retrieved. This circumstance also held true at higher frequencies, although some artifacts and ambiguities started to emerge. For instance, at 12 GHz, the OSM indicator was characterized by a hole at the center of the cylinder, while at 16 GHz, some artifacts appeared in the background. On the other hand, the B-IS indicator wrongly suggested some pixels inside the target, but such a circumstance did not compromise the correct estimation of the target support.

When a reduced amount of data was processed, both OSM and LSM did not work fine at the frequencies of 12 and 16 GHz, as shown in Figure 3. This was due to the fact that the amount of data, which was needed in order to correctly discretize the far field Equation (3), as well as to correctly compute the reduced scattered field Equation (5), was larger than M, whereas the size of the target became large in terms of wavelength. The B-IS was instead robust with respect to the reduction of data. In fact, the same performance (with respect to the case M = 36) was guaranteed when M = 18.

In order to quantitatively assess the quality of the reconstructions, the following dimensional error err_Δ was evaluated:

$$err_\Delta = \frac{r_{rec} - r_{ref}}{r_{ref}} \tag{8}$$

where r_{rec} and r_{ref} being the radius of the reconstructed and the reference cylinders, respectively. This metric allows one to appreciate the performance of the method in reconstructing the size of the target. In evaluating these errors, two different methods have been adopted to binarize the LSM and OSM indicator map. As the indicator maps are continuous functions, we first binarized them by applying the Canny edge detector [27,28] to its dB plot and set it to one of the pixels internal to the identified contour (see Table 2). The second method instead consisted of replacing all pixels in the input image with value greater than a given level $L \in \,]0,1]$ with the value 1 and replaced all other pixels with the value 0 (see Table 3). Note that the B-IS method did not require the adoption of a binarization technique. By comparing Tables 2 and 3, one can conclude that:

- the B-IS was more accurate in estimating the radius of the cylinder and it was robust with respect to the reduction of M;
- the choice of the thresholding techniques could impact the reconstruction of the support of the target. Indeed, the dimensional errors in Table 2 were different from the one in Table 3. In particular, the Canny edge detector-based approach led to an overestimation of the radius of the cylinder, while the second approach (L = 0.8) implied more accurate estimations, as witnessed by the lower dimensional errors. Notably, the accuracy of the reconstructions when LSM and OSM were adopted strongly depended on how the indicator map was binarized and on the adopted threshold.

Finally, in order to emphasize the flexibility of OSM and B-IS with respect to the measurement configuration, the inversion of a 72 × 9 data matrix was performed. In such a case the number of experiments was reduced to 9, while the measurements were increased to 72. The corresponding indicators are reported in Figure 4. B-IS was able to quite accurately retrieve the target support, while LSM only localized the target. Finally, differently from LSM, OSM was able to provide a rough reconstruction of the support, as it was sensitive with respect to the number of measurements rather than to the number of experiments.

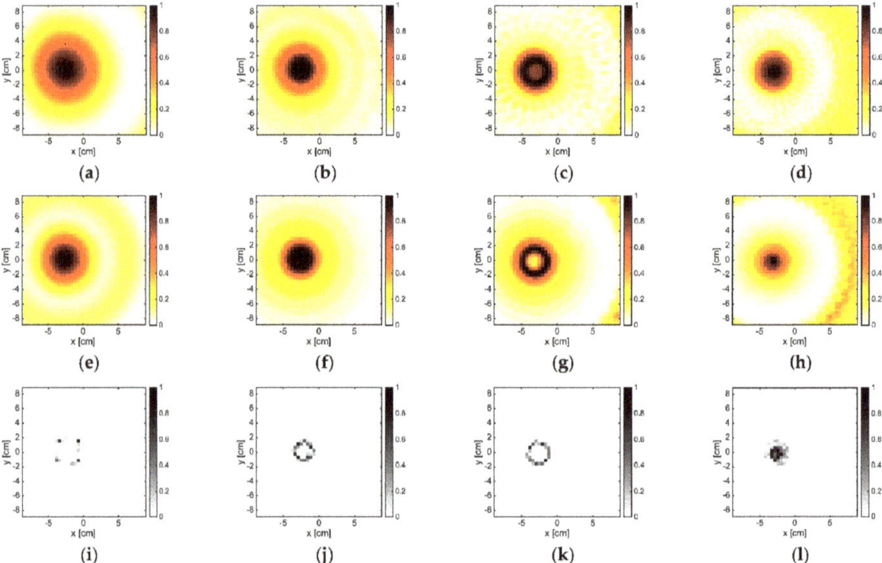

Figure 2. The DielTM target. From left to right: 2 GHz, 6 GHz, 12 GHz, and 16 GHz. From top to bottom: LSM, OSM, and B-IS indicators. The number of measurements and views was M = 36.

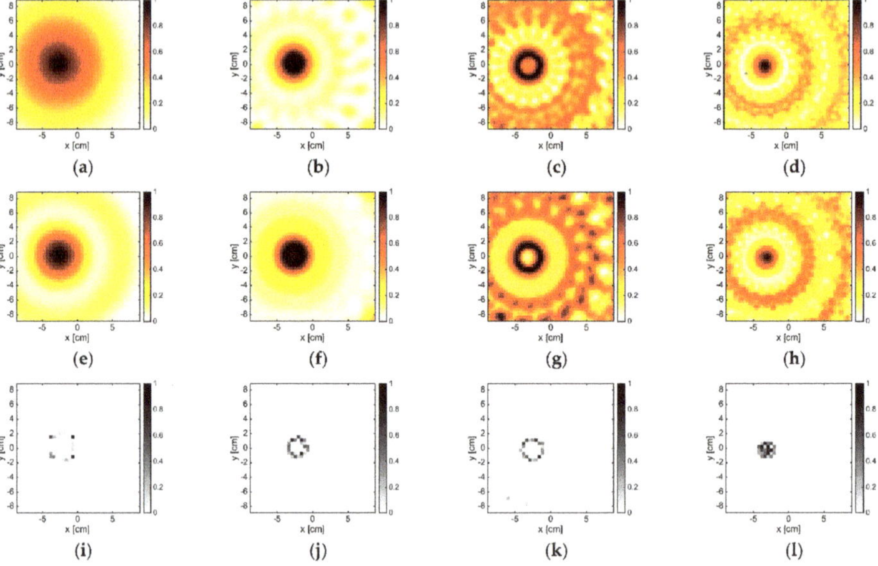

Figure 3. The DielTM target. From left to right: 2 GHz, 6 GHz, 12 GHz, and 16 GHz. From top to bottom: LSM, OSM, and B-IS indicators. The number of measurements and views was M = 18.

Table 2. Dimensional errors for the DielTM target. LSM and OSM indicators were binarized by applying the Canny edge detector. The symbol * indicates the cases wherein the indicator maps were not reliable, and the dimensional error was not computed.

Freq (GHz)	LSM		OSM		B-IS	
	M = 36	M = 18	M = 36	M = 18	M = 36	M = 18
2	1.4	0.47	0.97	0.97	0.18	0.18
6	0.55	0.55	0.69	0.74	0.18	0.11
12	1	*	1	*	0.18	0.18
16	1	*	0.18	*	0.18	0.11

Table 3. Dimensional errors for the DielTM target. LSM and OSM indicators were binarized by selecting L = 0.8. The symbol * indicates the cases wherein the indicator maps were not reliable, and the dimensional error was not computed.

Freq (GHz)	LSM		OSM		B-IS	
	M = 36	M = 18	M = 36	M = 18	M = 36	M = 18
2	0.37	0.32	0.11	0.11	0.18	0.18
6	0.22	0.09	0.27	0.28	0.18	0.11
12	0.47	*	0.47	*	0.18	0.18
16	0.03	*	0.47	*	0.18	0.11

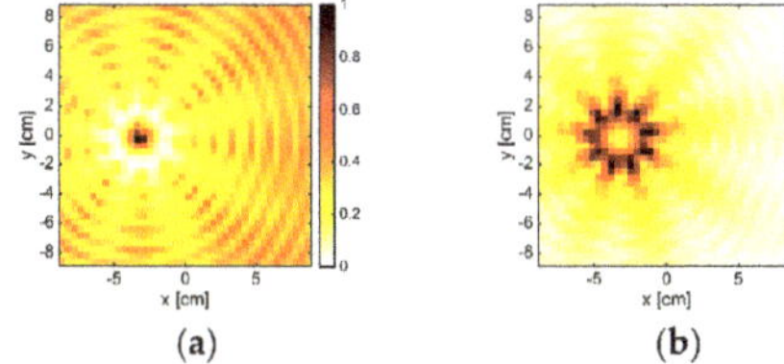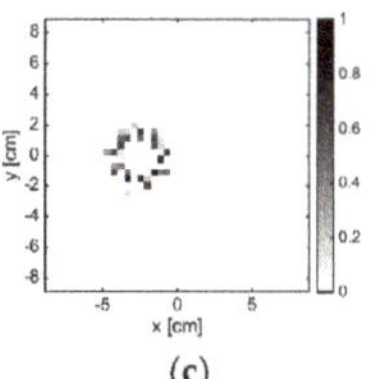

Figure 4. The DielTM target at 12 GHz. From left to right: LSM, OSM, and B-IS indicators. The dimension of the data matrix was 72 × 9.

3.2. Convex Metallic Target

Figure 5 depicts the reconstructions corresponding to the metallic rectangular targets pertaining to 2, 6, 10, and 16 GHz and M = 18. As can be observed, at the lowest frequency, the methods were just able to localize the scatterer, especially the B-IS method, which identified only some pixels in the center of the target. In fact, at 2 GHz, the target was just $0.2\lambda_b \times 0.1\lambda_b$ large. By increasing the frequency, the capability of the methods in retrieving the shape of the metallic target improved. For instance, at 6 GHz, both LSM and OSM correctly retrieved the unknown support, while B-IS overestimated its vertical dimension. However, in the frequency range of 8–16 GHz, both LSM and OSM came to exhibit some artifacts and did not retrieve an indicator map that was monotonically decreasing outside the actual support of the scatterer. This was correlated to the fact that the electromagnetic size of the target became larger and the amount of data needed to correctly solve Equation (3), as well as compute the reduced field Equation (5), was much larger than M. Indeed, by considering M = 36, reliable results, which are not shown for the sake of brevity, could be achieved. Notably, when M = 18, the B-IS was still able to reconstruct the actual support of the rectangular object also in the range of 8–16 GHz, as shown in Figure 5k,l.

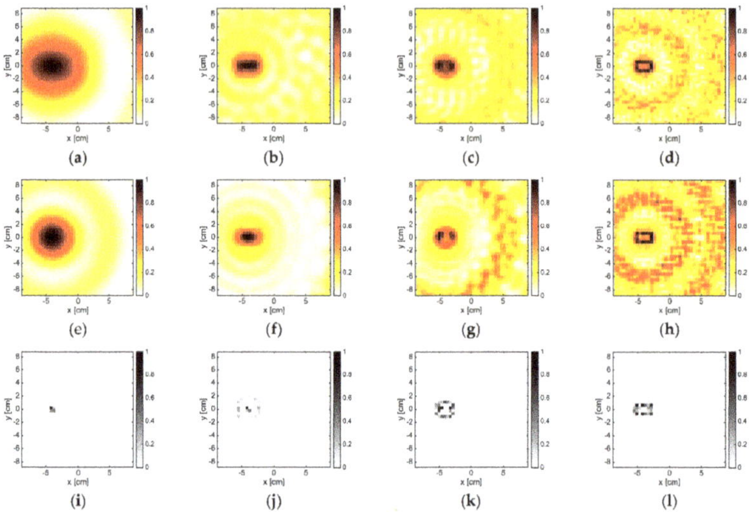

Figure 5. The RectTM_Dece target. From left to right: 2 GHz, 6 GHz, 10 GHz, and 16 GHz. From top to bottom: LSM, OSM, and B-IS indicators. The number of measurements and views was M = 18.

3.3. Non-Convex Metallic Target

The same behavior observed in the previous examples also holds true in case of the U-TM-shaped target (see Figure 6). In fact, at 2 GHz, the target was detected by all three methods and approximately localized by LSM and OSM, but only its convex-hull was retrieved by LSM and OSM. On the other hand, at higher frequencies, the support was only approximately retrieved by all three methods. In particular, some artifacts were present due to the mutual interactions between the "arms" of the target, especially in the LSM and OSM indicators, which were sensitive to the non-radiating part of the currents. Note that such a target belongs to the class of non-convex targets, so that it represents a more challenging case with respect to the previous ones.

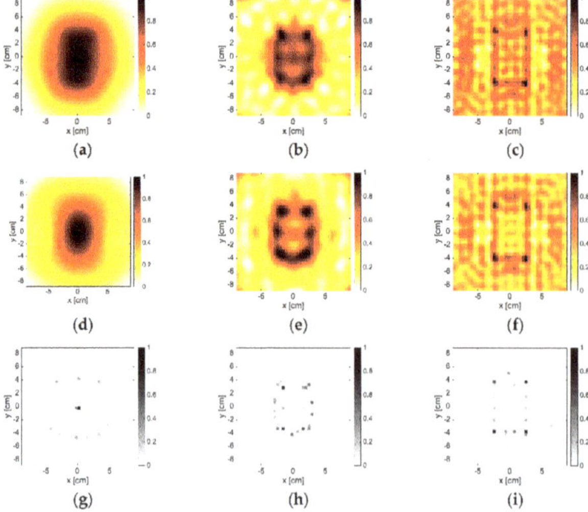

Figure 6. The U-TM-shaped target. From left to right: 2 GHz, 6 GHz, and 10 GHz. From top to bottom: LSM, OSM, and B-IS indicators. The number of measurements and views was M = 18.

No significant reconstructions were obtained by means of the three methods for M = 18 and a working frequency belonging to 12–16 GHz. In case of M = 36, the reconstructions are shown in Figure 7. As it can be seen, B-IS correctly retrieved the support of the target, apart from some spurious pixels, while the LSM did not provide reliable results due to the relevant presence of non-radiating currents and a high degree of non-linearity. Finally, the OSM indicator provided a rough support estimation only at 12 GHz.

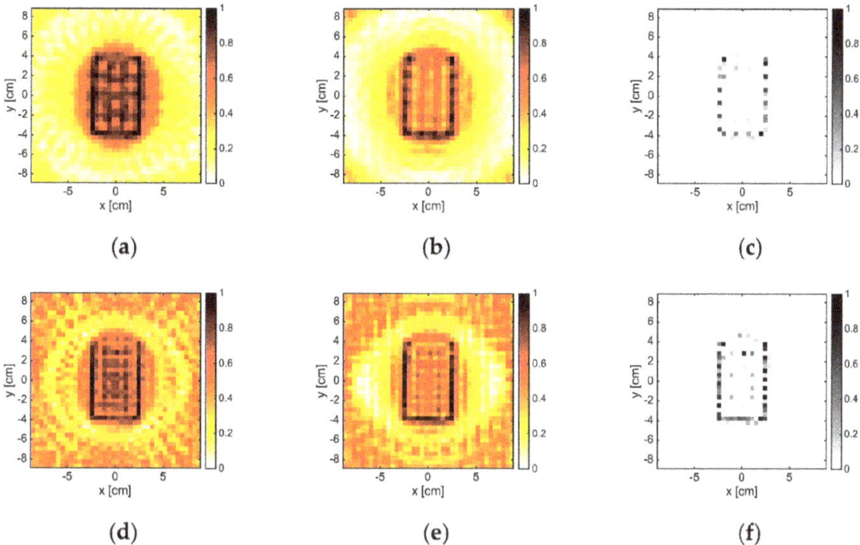

Figure 7. The U-TM-shaped target. From top to bottom: 12 GHz and 16 GHz. From left to right: LSM, OSM, and B-IS indicators. The number of measurements and views was 36.

3.4. Multiple Dielectric Targets

The results pertaining to the two identical cylinders are shown in Figure 8 for a number of experiments and measurements corresponding to M = 18. At 1 GHz, the methods could only detect the presence of an obstacle, but they could not identify the presence of two disjoint objects. On the contrary, at 2GHz, it was possible to distinguish two different targets; in particular, from the OSM map, one could also understand their shape. Note that the distance between the targets was 60 mm, which corresponded to $0.2\lambda_b$ at 1 GHz and $0.4\lambda_b$ at 2 GHz. When the frequency increased, both LSM and OSM worked fine. However, when the frequency was 6 GHz, some undesired artifacts occurred.

On the contrary, the B-IS method is able to provide quite accurate reconstructions of the support also in the frequency range 6–8 GHz. When the frequency and the size of the overall target's system were further increased, completely meaningless reconstructions were obtained by LSM and OSM, while B-IS indicator at least localized the two targets (see Figure 9g). In order to improve the reconstruction accuracy of the three methods, it was necessary to process a higher amount of data. This was witnessed by the reconstructions reported in Figure 9 corresponding to M = 36 at both 8 GHz and 16 GHz. However, at 16 GHz, the increase of the number of data was not enough for LSM and OSM to provide indicator maps free of artifacts.

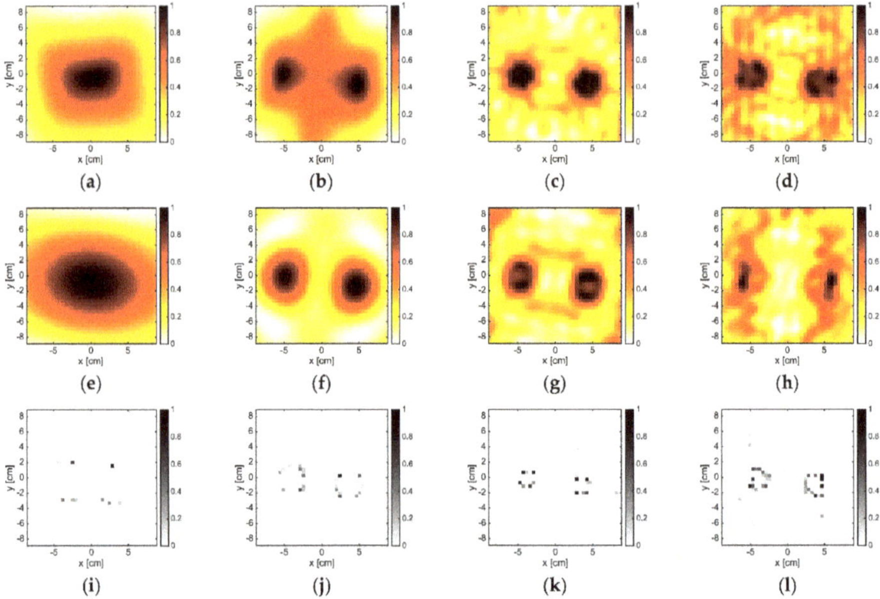

Figure 8. The TwinDielTM target. From left to right: 1 GHz, 2 GHz, 6 GHz, and 8 GHz. From top to bottom: LSM, OSM, and B-IS indicators. The number of measurements and views was M = 18.

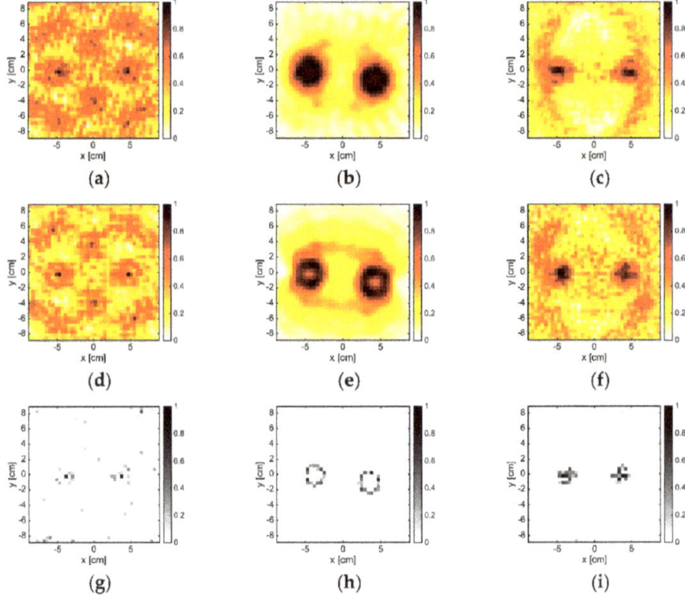

Figure 9. The TwinDielTM target. From left to right: 16 GHz (M = 18), 8 GHz (M = 36), and 16 GHz (M = 36). From top to bottom: LSM, OSM, and boundary indicators. The number of measurements and views was M = 18 for the first column and M = 36 for the last two columns.

4. Conclusions

The paper contributes to understand and discuss some strong and weak points of three qualitative methods for the inverse obstacle problem. In particular, a comparison among the popular LSM, the OSM, and the more recent B-IS was reported by considering the processing of the 2001 Fresnel dataset, which included both dielectric and metallic targets, as well as non-convex and multiple targets. Different frequencies and amount of data were considered for each method. In all cases, because of the characteristics of the Fresnel measurements setup, data were aspect-limited, i.e., not all incidence observation angle couples were available (for instance the monostatic data was missing). Also, scattered field data turned out to be undersampled at the higher frequencies with respect to the criteria (depending on the dimension of the investigation domain) discussed in Reference [26].

Results suggested that at lower frequencies, reliable reconstructions were guaranteed if the size of the target was higher than $0.2\lambda_b$. Indeed, for lower dimensions, the three methods could only localize the target, while its size was overestimated. Such a result was fully consistent with expected limitations on resolution implied by the diffraction theorem [29] and the results in Reference [26]. When the frequency was much larger, LSM and OSM began to exhibit some artifacts and, if the amount of data was not sufficient or due to the presence of non-negligible non-radiating sources, they could provide unreliable reconstructions. Notably, the B-IS seemed to be more robust with respect to LSM and OSM when undersampling data and/or increasing the frequency. In fact, it could still provide accurate indicator maps in a number of cases where LSM and OSM failed.

The examples have also proved that both OSM and LSM were sensitive to the choice of the binarization technique to discriminate between points inside and outside the scatterer. On the other hand, both OSM and LSM had a negligible computational burden, while B-IS had to solve a CP problem whose computational burden increased with the cells used to discretize D, as well as with M. In the considered examples, the computational time amounted to a few seconds for OSM and LSM, and to a few minutes for B-IS.

Future work will be devoted to extending such a performance comparison to other qualitative methods, for instance, the modified version of LSM [30,31].

Author Contributions: The authors contribute equally.

Funding: This work was partly supported by the Italian Ministry of Education, Universities, and Research through the Project of National Interest (PRIN) 'Field and Temperature Shaping for MWI Hyperthermia - FAT SAMMY' under Grant 2015KJE87K.

Acknowledgments: The authors want to acknowledge Prof. Tommaso Isernia for the fruitful discussions, which inspired the writing of this paper.

References

1. Colton, D.; Kress, R. *Inverse Acoustic and Electromagnetic Scattering Theory*, 2nd ed.; Springer: Berlin, Germany, 1998; ISBN 9781461449423.
2. Pastorino, M. *Microwave Imaging*; John Wiley: New York, NY, USA, 2010.
3. Conceicção, R.C.; Mohr, J.J.; O'Halloran, M. (Eds.) *An Introduction to Microwave Imaging for Breast Cancer Detection*, 1st ed.; Biological and Medical Physics, Biomedical Engineering; Springer International Publishing: Basel, Switzerland, 2016.
4. Crocco, L.; Conceicção, R.C.; James, M.L.; Karanasiou, I. (Eds.) *Emerging Electromagnetic Technologies for Brain Diseases Diagnostics, Monitoring and Therapy*; Springer: Cham, Switzerland, 2018.
5. Bevacqua, M.T.; Bellizzi, G.G.; Crocco, L.; Isernia, T. A Method for Quantitative Imaging of Electrical Properties of Human Tissues from Only Amplitude Electromagnetic Data. *Inverse Probl.* **2019**, *35*, 025006. [CrossRef]
6. Turk, A.S.; Hocaoglu, A.K.; Vertiy, A.A. (Eds.) *Subsurface Sensing*; Wiley: Hoboken, NJ, USA, 2011.
7. Persico, R. *Introduction to Ground Penetrating Radar: Inverse Scattering and Data Processing*; Wiley: Hoboken, NJ, USA, 2014.

8.	Bertero, M.; Boccacci, P. *Introduction to Inverse Problems in Imaging*; Institute of Physics: Bristol, UK, 1998.

9.	Cakoni, F.; Colton, D. *Qualitative Methods in Inverse Scattering Theory*; Springer: Berlin, Germany, 2006.

10.	Ammari, H.; Iakovleva, E.; Lesselier, D.; Perrusson, G. MUSIC-type electromagnetic imaging of a collection of small three-dimensional inclusions. *Siam J. Sci. Comput.* **2007**, *29*, 674–709. [CrossRef]

11.	Tortel, H.; Micolau, G.; Saillard, M. Decomposition of the time reversal operator for electromagnetic scattering. *J. Electromagn. Waves Appl.* **1999**, *13*, 687–719. [CrossRef]

12.	Colton, D.; Haddar, H.; Piana, M. The linear sampling method in inverse electromagnetic scattering theory. *Inverse Probl.* **2003**, *19*, 105–137. [CrossRef]

13.	Kirsch, A.; Grinberg, N.I. *The Factorization Method for Inverse Problems*; Oxford University Press: Oxford, UK, 2008.

14.	Potthast, R. A study on orthogonality sampling. *Inverse Probl.* **2010**, *26*, 074015. [CrossRef]

15.	Bevacqua, M.; Isernia, T. Shape reconstruction via equivalence principles, constrained inverse source problems and sparsity promotion. *Prog. Electromagn. Res.* **2017**, *158*, 37–48. [CrossRef]

16.	Bevacqua, M.T.; Isernia, T. Boundary Indicator for Aspect Limited Sensing of Hidden Dielectric Objects. *IEEE Geosci. Remote Sens. Lett.* **2018**, *15*, 838–842. [CrossRef]

17.	Donoho, D. Compressed sensing. *IEEE Trans. Inf. Theory* **2006**, *52*, 1289–1306. [CrossRef]

18.	Franceschetti, G. *Electromagnetics: Theory, Techniques, and Engineering Paradigms*; Springer Science & Business Media: Berlin, Germany, 2013.

19.	Belkebir, K.; Saillard, M. Special section: Testing inversion algorithms against experimental data. *Inverse Probl.* **2001**, *17*, 1565–2028. [CrossRef]

20.	Catapano, I.; Crocco, L.; Isernia, T. On simple methods for shape reconstruction of unknown scatterers. *IEEE Trans. Antennas Propag.* **2007**, *55*, 1431–1436. [CrossRef]

21.	Kowalski, M. Sparse regression using mixed norms. *Appl. Comput. Harmon. Anal.* **2009**, *27*, 303–324. [CrossRef]

22.	Fornasier, M.; Rauhut, H. Recovery algorithms for vector-valued data with joint sparsity constraints. *SIAM J. Numer. Anal.* **2008**, *46*, 577–613. [CrossRef]

23.	Di Donato, L.; Bevacqua, M.; Isernia, T.; Catapano, I.; Crocco, L. Improved quantitative microwave tomography by exploiting the physical meaning of the Linear Sampling Method. In Proceedings of the 5th European Conference on Antennas and Propagation, Rome, Italy, 11–15 April 2011; pp. 3828–3831.

24.	Bevacqua, M.T.; Palmeri, R.; Isernia, T.; Crocco, L. Physical Interpretation of the Orthogonality Sampling Method. In Proceedings of the 2nd URSI Atlantic Radio Science Meeting (AT-RASC), Gran Canaria, Spain, 28 May–1 June 2018.

25.	Di Donato, L.; Bevacqua, M.T.; Crocco, L.; Isernia, T. Inverse Scattering Via Virtual Experiments and Contrast Source Regularization. *IEEE Trans. Antennas Propag.* **2015**, *63*, 1669–1677. [CrossRef]

26.	Bucci, O.M.; Isernia, T. Electromagnetic inverse scattering: Retrievable information and measurement strategies. *Radio Sci.* **1997**, *32*, 2123–2138. [CrossRef]

27.	Canny, J. A computational approach to edge detection. *IEEE Trans Pattern Anal. Mach. Intell.* **1986**, *8*, 679–698. [CrossRef] [PubMed]

28.	Catapano, I.; Crocco, L.; D'Urso, M.; Isernia, T. On the Effect of Support Estimation and of a New Model in 2-D Inverse Scattering Problems. *IEEE Trans. Antennas Propag.* **2007**, *55*, 1895–1899. [CrossRef]

29.	Devaney, A.J. *Mathematical Foundations of Imaging, Tomography and Wavefield Inversion*; Cambridge University Press: Cambridge, UK, 2012.

30.	Crocco, L.; Di Donato, L.; Catapano, I.; Isernia, T. An Improved Simple Method for Imaging the Shape of Complex Targets. *IEEE Trans. Antennas Propag.* **2013**, *61*, 843–851. [CrossRef]

31.	Agarwal, K.; Chen, X.; Zhong, Y. A multipole-expansion based linear sampling method for solving inverse scattering problems. *Opt. Express* **2010**, *18*, 6366–6381. [CrossRef] [PubMed]

Article

A Finite-Difference Approach for Plasma Microwave Imaging Profilometry

Loreto Di Donato [1,2,*], **David Mascali** [2], **Andrea F. Morabito** [3] **and Gino Sorbello** [1,2]

[1] Department of Electrical, Electronics and Computer Engineering (DIEEI), University of Catania, viale A. Doria 6, 95126 Catania, Italy
[2] Laboratori Nazionali del Sud, National Institute of Nuclear Physics (INFN), Via Santa Sofia 62, 95125 Catania, Italy
[3] Department of Information Engineering, Infrastructures and Sustainable Energy (DIIES), University "Mediterranea" of Reggio Calabria, via Graziella, Loc. Feo di Vito, 89124 Reggio di Calabria, Italy
* Correspondence: loreto.didonato@dieei.unict.it

Received: 9 July 2019; Accepted: 9 August 2019; Published: 12 August 2019

Abstract: Plasma diagnostics is a topic of great interest in the physics and engineering community because the monitoring of plasma parameters plays a fundamental role in the development and optimization of plasma reactors. Towards this aim, microwave diagnostics, such as reflectometric, interferometric, and polarimetric techniques, can represent effective means. Besides the above, microwave imaging profilometry (MIP) may allow the obtaining of tomographic, i.e., volumetric, information of plasma that could overcome some intrinsic limitations of the standard non-invasive diagnostic approaches. However, pursuing MIP is not an easy task due to plasma's electromagnetic features, which strongly depend on the working frequency, angle of incidence, polarization, etc., as well as on the need for making diagnostics in both large (meter-sized) and small (centimeter-sized) reactors. Furthermore, these latter represent extremely harsh environments, wherein different systems and equipment need to coexist to guarantee their functionality. Specifically, MIP entails solution of an inverse scattering problem, which is non-linear and ill-posed, and, in addition, in the one-dimensional case, is also severely limited in terms of achievable reconstruction accuracy and resolution. In this contribution, we address microwave inverse profiling of plasma assuming a high-frequency probing regime when magnetically confined plasma can be approximated as both an isotropic and weak penetrable medium. To this aim, we adopt a finite-difference frequency-domain (FDFD) formulation which allows dealing with non-homogeneous backgrounds introduced by unavoidable presence of plasma reactors.

Keywords: microwave plasma diagnostics; electromagnetic inverse scattering; microwave imaging profilometry; finite-difference methods

---

1. Introduction

Plasma diagnostics using active microwave techniques is one of the most promising tools to optimize the production of plasma in both large and compact reactors because of its non-invasive nature and relatively modest access requirements [1]. Among the others, microwave imaging reflectometry (MIR), interferometric and polarimetric techniques are currently under investigation to extract plasma proprieties useful to model and optimize the heating process. Such information is crucial to study and improve plasma generation based on Electron Cyclotron Resonance [2] as well as innovative heating schemes such as the ones employing Electrostatic Bernstein Waves in overdense plasma [3]. Unfortunately, the diagnostic approaches mentioned above show some limitations. As an example, MIR, a multifrequency radar-like technique, mainly used in large reactors, such as tokamak and stellarator, is commonly used to infer electronic density distribution and fluctuation, but suffers from the impossibility of performing diagnostics of non-increasing plasma profiles since inner reflecting cutoff layers are "hidden" from the outer ones [4]. On the other hand, line-integrated measurements, such as those actually provided by interferometric and polarimetric techniques, although employable on compact reactors such as the electron cyclotron ion source (ECRIS) [5], cannot give local information about electron density, but only a line-of-sight integrated value.

For these reasons, it is worth devoting interest towards microwave imaging tomographic approaches, such as those actually used in medical and subsurface microwave diagnostics [6], which are in principle able to "locally" monitor plasma uniformity and instabilities.

In a recent paper, we investigated the possibility of obtaining plasma imaging profilometry (MIP) by means of electromagnetic inverse scattering techniques requiring only measurements of the reflection coefficient in a quite large frequency band. In particular, adopting a high-frequency probing regime (i.e., probing frequencies much larger than the cyclotron frequency), i.e., when the plasma can be assumed to be both isotropic and weak scatterer, linear imaging recovery techniques are able to achieve quantitative characterization of plasma profiles [7]. The results have shown that if the recovery process is properly formulated as a sparse optimization problem, compressive sensing (CS)-based strategies are able to achieve nearly optimal results. However, free-space homogeneous background has been considered, which is an assumption suitable only if multiple reflections introduced by the metallic environment (reactor walls) are neglected or, in any case, filtered out.

With the aim to get a further step in the analysis of such a problem, in this paper we introduce a frequency-domain finite-difference (FDFD) formulation for the solution of the inverse scattering problem in order to take easily into account the effect of non-homogeneous backgrounds related to the metallic environment of the plasma chamber.

On the other hand, besides the modeling aspects, solution of the inverse scattering problem in presence of a metallic surface, or, in general, non-homogeneous background, can bring some advantages due to the possibility to achieve enhanced resolution in the reconstruction process [8,9]. Roughly speaking, this enhancement can be explained due to the multiple reflected waves which can bring additional information conveyed back by the recovery process. Interestingly, this may be of fundamental importance in the solution of the one-dimensional inverse scattering problem (also known as inverse profiling) wherein the number and kind of retrievable information is severely limited [10].

The paper is structured as follows. In Section 2, the adopted mathematical formulation for the solution of the forward and inverse scattering problem is described. In Section 3 the microwave imaging profilometry for inhomogeneous plasma slabs recovery is formulated. Finally, in Section 4, the developed approach is validated against simulated data dealing with homogeneous and non-homogeneous backgrounds as well as reflection-only and reflection-transmission measurement configuration. Conclusions follow.

2. Mathematical Formulation

Let us assume a bounded, simply connected, domain D located in an non-homogeneous background medium, wherein some electromagnetic primary sources $\mathbf{j}_0$ are located in a region characterized by relative permittivity and permeability distribution $\mu_r(\mathbf{r}')$ and $\epsilon_r(\mathbf{r}')$, $\mathbf{r}'$ spanning the space D. After normalizing the electric field, magnetic field and the primary sources according to $\tilde{\mathbf{e}} = \sqrt{\frac{\epsilon_0}{\mu_0}}\underline{\mathbf{e}}, \tilde{\mathbf{h}} = \underline{\mathbf{h}}$ and $\tilde{\mathbf{j}}_0 = c_0\mathbf{j}_0$, with $c_0 = \dfrac{1}{\sqrt{\epsilon_0\mu_0}}$ the speed of light, Maxwell equations read:

$$c_0\underline{\nabla} \times \tilde{\mathbf{e}} = -j\omega\mu_r(\mathbf{r}')\tilde{\mathbf{h}}$$

$$c_0\underline{\nabla} \times \tilde{\mathbf{h}} = j\omega\epsilon_r(\mathbf{r}')\tilde{\mathbf{e}} + \tilde{\mathbf{j}}_0$$

(1)

where ω is the angular frequency and the time factor $\exp\{j\omega t\}$ is assumed and dropped for the time-harmonic fields. Normalized electric and magnetic fields, $\tilde{\mathbf{e}}$ and $\underline{\mathbf{h}}$, are of the same order of magnitude and this is an advantage in formulating the perfectly matched layer (PML) [11]. If, for the sake of a simpler notation, we drop the tilde ˜ symbol and the c_0 is included in the $\underline{\nabla} \times$ definition, we can write:

$$\tilde{\underline{\nabla}} \times \underline{e} = -j\omega \mu_r(\underline{r}')\underline{h}$$

$$\tilde{\underline{\nabla}} \times \underline{h} = j\omega \epsilon_r(\underline{r}')\underline{e} + \underline{j}_0 \tag{2}$$

In the above formulation, to satisfy different boundary conditions, a lossy PML [12] can be introduced to truncate the computational domain, as commonly done in numerical solver. By choosing proper PML parameters, various boundary conditions can seamlessly introducing ranging from a perfect absorbing boundary condition to a PEC (or PMC) boundary condition, the latter case when no PML is used [11].

The background field (i.e., the field without the unknown profile) defined as a plane wave propagating in a non-homogeneous, eventually lossy, background with permittivity and permeability, respectively, given by:

$$\epsilon_b(\underline{r}') = \epsilon_b'(\underline{r}') - j\epsilon_b''(\underline{r}') = \epsilon_b'(\underline{r}') - j\frac{\sigma(\underline{r}')}{\epsilon_0 \omega}$$

$$\mu_b(\underline{r}') = \mu_b'(\underline{r}') - j\mu_b''(\underline{r}')$$

satisfies the following equations:

$$\tilde{\underline{\nabla}} \times \underline{e}_{bck} = -j\omega \mu_b(\underline{r}')\underline{h}_{bck}$$

$$\tilde{\underline{\nabla}} \times \underline{h}_{bck} = j\omega \epsilon_b(\underline{r}')\underline{e}_{bck} + \underline{j}_0 \tag{3}$$

To formulate the scattering problem, we can add and subtract the quantities $j\omega \epsilon_b(\underline{r}')\underline{e}$ and $j\omega \mu_b(\underline{r}')\underline{h}$ at the right hand side member of Equation (2) to obtain:

$$\tilde{\underline{\nabla}} \times \underline{e} = -j\omega \mu_b(\underline{r}')\underline{h} - j\omega \left[\mu_r(\underline{r}') - \mu_b(\underline{r}')\right]\underline{h}$$

$$\tilde{\underline{\nabla}} \times \underline{h} = j\omega \epsilon_b \underline{e} + j\omega \left[\epsilon_r(\underline{r}') - \epsilon_b(\underline{r}')\right]\underline{e} + \underline{j}_0 \tag{4}$$

and, finally, subtracting (3) to (4), the equations governing the scattered fields, defined as $\underline{e}_{sct} = \underline{e} - \underline{e}_{bck}$ and $\underline{h}_{sct} = \underline{h} - \underline{h}_{bck}$, are given by:

$$\tilde{\underline{\nabla}} \times \underline{e}_{sct} = -j\omega \mu_b(\underline{r}')\underline{h}_{sct} - j\omega \mu_b(\underline{r}') \left[\frac{\mu_r(\underline{r}') - \mu_b(\underline{r}')}{\mu_b(\underline{r}')}\right]\underline{h}$$

$$\tilde{\underline{\nabla}} \times \underline{h}_{sct} = j\omega \epsilon_b(\underline{r}')\underline{e}_{sct} + j\omega \epsilon_b(\underline{r}') \left[\frac{\epsilon_r(\underline{r}') - \epsilon_b(\underline{r}')}{\epsilon_b(\underline{r}')}\right]\underline{e} \tag{5}$$

wherein the quantities under the square brackets can be easily recognized as the contrast functions:

$$\chi_\mu(\omega, \underline{r}') = \frac{\mu_r(\omega, \underline{r}') - \mu_b(\omega, \underline{r}')}{\mu_b(\omega, \underline{r}')},$$

$$\chi_\epsilon(\omega, \underline{r}') = \frac{\epsilon_r(\omega, \underline{r}') - \epsilon_b(\omega, \underline{r}')}{\epsilon_b(\omega, \underline{r}')},$$

which relate the dielectric and magnetic properties of the anomalies, i.e., scattering objects, to those of the embedding background medium at each frequency.

Adopting a dual grid [13,14] and a proper discretization, in operator notation, the above equations can be rewritten as:

$$\begin{bmatrix} \mathcal{C} & j\omega \mathcal{M}_b \\ -j\omega \mathcal{E}_b & \mathcal{C} \end{bmatrix} \begin{bmatrix} \underline{e}'_{sct} \\ \underline{h}'_{sct} \end{bmatrix} = \begin{bmatrix} 0 & -j\omega \mu_b \chi_\mu \\ j\omega \epsilon_b \chi_\epsilon & 0 \end{bmatrix} \begin{bmatrix} \underline{e}' \\ \underline{h}' \end{bmatrix}, \tag{6}$$

wherein $()'$ is used to indicate the field integrated over the appropriate line element of the primal or dual mesh ($\underline{e}' \approx \underline{e}\overline{\Delta L}$, $\underline{h}' \approx \underline{h}\widehat{\Delta L}$). Moreover, the matrix operator C known in the literature as incidence matrix (see for example section §7.3.3 and Equation (7.1) of Tonti's book [13] or [15]) denotes the proper derivative operations and is a sparse matrix. Moreover, $\mathcal{M}_b$ and $\mathcal{E}_b$ are matrices accounting for the magnetic and dielectric non-homogeneous background properties, respectively, wherein $\mathcal{E}_b \approx \epsilon_b \frac{\widehat{\Delta S}}{\overline{\Delta L}}$ and $\mathcal{M}_b \approx \mu_b \frac{\overline{\Delta S}}{\widehat{\Delta L}}$, with $\overline{\Delta L}$ and $\overline{\Delta S}$ refer to lines and surfaces of the primal mesh, and $\widehat{\Delta L}$ and $\widehat{\Delta S}$ refer to the dual mesh [13]. In the remaining of the paper for the sake of a simpler notation we drop the $'$ symbol.

The formal solution of the linear system (6) can be rewritten as:

$$
\begin{bmatrix} \underline{e}_{sct} \\ \underline{h}_{sct} \end{bmatrix} = \mathcal{H}_b^{-1} \left(\begin{bmatrix} 0 & -j\omega\mu_b\chi_\mu \\ j\omega\epsilon_b\chi_\epsilon & 0 \end{bmatrix} \begin{bmatrix} \underline{e} \\ \underline{h} \end{bmatrix} \right). \tag{7}
$$

The above equation is the key equation for the inverse scattering problem adopting a finite-difference formulation. As a matter of fact, the measured scattered fields, i.e., the scattered field recorded on the data domain S (for the case at hand, the same positions $z = z_S$ at different frequencies), can be represented by the following operator notations:

$$
\begin{bmatrix} \underline{e}_{sct}(z_S, \omega_n) \\ \underline{h}_{sct}(z_S, \omega_n) \end{bmatrix} = \mathcal{M}_S \left\{ \mathcal{H}_b^{-1} \left(\begin{bmatrix} 0 & -j\omega\mu_b\chi_\mu \\ j\omega\epsilon_b\chi_\epsilon & 0 \end{bmatrix} \begin{bmatrix} \underline{e} \\ \underline{h} \end{bmatrix} \right) \right\}, \tag{8}
$$

where $\mathcal{M}_S$ is an operator that extracts the field values at the observation point $z = z_S$ at different frequencies. Substituting $[\underline{e}_{sct} , \underline{h}_{sct}]^T = [\underline{e} , \underline{h}]^T - [\underline{e}_{bck} , \underline{h}_{bck}]^T$ into (7), we obtain the domain (or object) equation:

$$
\begin{bmatrix} \underline{e}_{bck} \\ \underline{h}_{bck} \end{bmatrix} = \begin{bmatrix} \underline{e} \\ \underline{h} \end{bmatrix} - \left\{ \mathcal{H}_b^{-1} \left(\begin{bmatrix} 0 & -j\mu_b\omega\chi_\mu \\ j\omega\epsilon_b\chi_\epsilon & 0 \end{bmatrix} \begin{bmatrix} \underline{e} \\ \underline{h} \end{bmatrix} \right) \right\}, \tag{9}
$$

that can be sampled with an operator, say it $\mathcal{M}_D$, that extracts fields only in the imaging domain D at the discretization grid points z_D [16], i.e., :

$$
\begin{bmatrix} \underline{e}_{bck}(z_D, \omega_n) \\ \underline{h}_{bck}(z_D, \omega_n) \end{bmatrix} - \begin{bmatrix} \underline{e}(z_D, \omega_n) \\ \underline{h}(z_D, \omega_n) \end{bmatrix} = \mathcal{M}_D \left\{ \mathcal{H}_b^{-1} \left(\begin{bmatrix} 0 & -j\omega\mu_b\chi_\mu \\ j\omega\epsilon_b\chi_\epsilon & 0 \end{bmatrix} \begin{bmatrix} \underline{e} \\ \underline{h} \end{bmatrix} \right) \right\}. \tag{10}
$$

Equations (8)–(10) are the basic equations to solve the inverse scattering problem which is non-linear since both the contrast functions χ_ϵ and χ_μ and the field $\underline{e}$ and $\underline{h}$ are unknowns of the problem [17]. Moreover, the problem is ill-posed, too, and effective regularization strategies must be adopted for stable solutions against the presence of noise on data [18].

3. Microwave Imaging Profilometry

3.1. Linearized Approach for Plasma Slabs

Let us consider a one-dimensional non-homogeneous slab extending in the range $[z_1, z_2]$ embedded in a non-homogeneous background, see Figure 1a. Normal incidence plane waves are used to probe the slab at difference frequencies with phase reference equal to zero at $z = z^{(\text{View})}(z \geq z_1)$. Accordingly, the scattered field can be gathered at a given abscissa $z_1 \leq z_S = z^{(\text{Meas})} \leq z_{min}$ (reflection measurement) or $z_{max} \leq z_S = z^{(\text{Meas})} \leq z_2$ (transmission measurement).

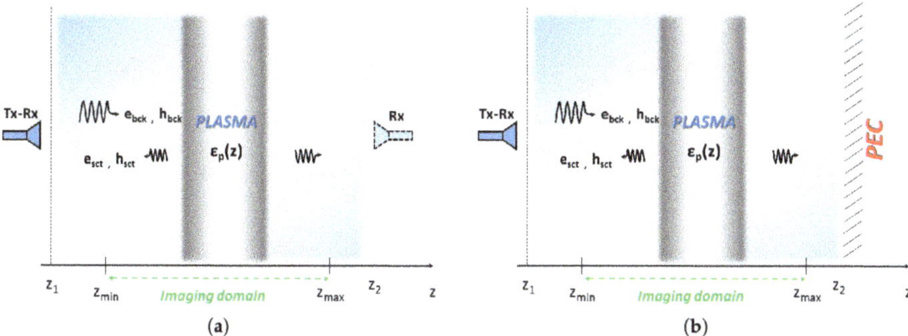

Figure 1. Sketch of the single view normal incidence measurement. The imaging domain extents in the range $[z_{min}, z_{max}]$ and the scattered field data at different frequencies are gathered at a given observation points external to this range. (**a**) Homogeneous background: configuration with only an active (tx-rx) antenna (neglecting the rightmost dashed antenna) and an active (tx-rx) and a passive (rx) antenna (considering the rightmost dashed antenna). (**b**) Non-homogeneous background: configuration with an active antenna (tx-rx) with a PEC surface.

To retrieve nonmagnetic ($\mu_r(\mathbf{r}') = \mu_b$) dielectric profile of inhomogeneous isotropic plasma slabs, we consider a first order linearized approach under the Rytov approximation, instead of solving the exact optimization problems stated by Equations (8)–(10). Linearized approaches, such as the Born (BA) and Rytov (RA) approximations, can be used to recover small and extended (with respect to the wavelength) weak slabs, respectively, i.e., slabs showing low contrast ($\|\chi_\epsilon\| << 1$). In particular, the Rytov approximation is useful for scatterers far extended in terms of the probing wavelengths (anyway, RA reduces to BA when the scatterers become smaller and smaller with respect to the probing wavelengths [19]). In this respect, to fulfill model hypothesis, it is worth noticing that assuming a relatively high-frequency probing band (with respect to the cyclotron frequency) entails several positive fallouts. First, at probing frequencies much higher than the cyclotron frequency, the plasma can be assumed as an isotropic medium since the permittivity tensor becomes diagonal with the terms all equals each other [2,20]. Second, at probing frequencies such that $\omega^2 >> \nu^2$ the plasma contrast becomes:

$$\chi_\epsilon = -\frac{\omega_p^2}{\omega^2 + \nu^2} - j\frac{\omega_p^2 \nu}{\omega(\omega^2 + \nu^2)} \approx -\frac{\omega_p^2}{\omega^2} - j\frac{\omega_p^2 \nu}{\omega^3}, \tag{11}$$

wherein:

$$\omega_p = \sqrt{\frac{n_e e^2}{m_e \epsilon_0}}, \tag{12}$$

n_e, e and m_e being the electron density, electron charge, and electron mass, respectively. Under the above approximations, considering the frequency difference (FD) formulation described in Section 2, the unknown total electric field can be approximated by the incident field according to the RA, to obtain the following linear equation:

$$\mathbf{f}_{sct} = \mathcal{M}_S\left\{\tilde{\mathcal{R}}_b^{-1}\chi_\epsilon \mathbf{e}_{bck}\right\}, \tag{13}$$

wherein $\mathbf{f}_{sct} = \mathbf{e}_{sct}/\mathbf{e}_{bck}$ and $\tilde{\mathcal{R}}_b^{-1}$, replacing $\tilde{\mathcal{H}}_b^{-1}$ unless unessential constant, takes into account the normalization by the incident field values at the measurement point $z^{(Meas)}$ introduced by the Rytov approximation [19] and the frequency dependence of the (multifrequency) scattering operator. As a result, in order to solve for the plasma constitutive parameters ω_p^2 and ν, the frequency dependent complex valued problem (13), can be recast in terms of frequency independent real valued problem as already exploited in [7], i.e.,:

$$
\begin{bmatrix} Re(\mathbf{f}_{sct}) \\[2mm] Im(\mathbf{f}_{sct}) \end{bmatrix} = \begin{bmatrix} -\frac{1}{\omega^2}\mathcal{M}_S Re(\widetilde{\mathcal{R}}_b^{-1}\mathbf{e}_{bck}) & \frac{1}{\omega^3}\mathcal{M}_S Im(\widetilde{\mathcal{R}}_b^{-1}\mathbf{e}_{bck}) \\[2mm] -\frac{1}{\omega^3}\mathcal{M}_S Im(\widetilde{\mathcal{R}}_b^{-1}\mathbf{e}_{bck}) & -\frac{1}{\omega^2}\mathcal{M}_S Re(\widetilde{\mathcal{R}}_b^{-1}\mathbf{e}_{bck}) \end{bmatrix} \begin{bmatrix} \omega_p^2 \\[2mm] \omega_p^2\nu \end{bmatrix}. \tag{14}
$$

Problem (14) can be solved as a whole or by splitting it into two subsequent subproblems wherein the first step copes with the retrieval of the plasma frequency (i.e., the real part of the plasma contrast) and the second one dealing with the retrieval of the collision rate which is related to the imaginary part of the plasma contrast. This approach has been proposed in [7] wherein, considering the very low effect of the plasma losses on scattered field, the overall problem can be "decoupled". This notwithstanding, in the following, we concentrate our attention only on the retrieval of the plasma angular frequency. As a matter of fact, experimental studies state that the collision rate is almost one order of magnitude smaller that the cyclotron frequency, which in turn, is much lower that the probing frequency. This entails to cope with a lossless plasma and the whole entire problem (14) will be reduced to the following one:

$$
\begin{bmatrix} Re(\mathbf{f}_{sct}) \\[2mm] Im(\mathbf{f}_{sct}) \end{bmatrix} = \begin{bmatrix} -\omega_{max}^2\frac{\omega}{\omega^2}\mathcal{M}_S Re(\widetilde{\mathcal{R}}_b^{-1}\mathbf{e}_{bck}) \\[2mm] -\omega_{max}^2\frac{\omega}{\omega^3}\mathcal{M}_S Im(\widetilde{\mathcal{R}}_b^{-1}\mathbf{e}_{bck}) \end{bmatrix} \begin{bmatrix} \frac{\omega_p^2}{\omega_{max}^2} \end{bmatrix}, \tag{15}
$$

with ω_{max}^2 the maximum probing angular frequency used as normalization constant introduced for numerical stability of the inversion process [7].

3.2. Sparsity-Promoting Recovery Approaches

Equation (15) is a linear but still ill-posed one which must be solved in a regularized fashion. However, as stated in [10] and investigated in the case of plasma [7], minimum energy solutions approaches, such as Tikonov regularization [21,22] are completely unable to recover neither the shape nor the constitutive parameters of arbitrarily shaped plasma slabs. For this reason, CS-based approaches are worth to be considered, since they are in principle suitable to overcome the intrinsic limitations of the minimum energy solution related to the kind and number of actual parameters which can be conveyed back from multifrequency single view inverse scattering problem.

Let us consider a generic linear problem of the following kind:

$$
\Phi x = f, \tag{16}
$$

x is the $N \times 1$ unknown vector and Φ is the $M \times N$ matrix, the so-called sensing matrix, and f the $M \times 1$ data vector of the multifrequency scattered field ordered according to (15). Let us suppose that a convenient representation matrix Ψ exists such that the unknown can be expanded as $x = \Psi s$ with only a small number of the coefficients s different from zero. According to CS theory, it is possible to solve the inverse problem (16) even if $M << N$ but is sufficiently larger than the number S of coefficients different from zero (with $S < M < N$). However, it is worth underlining that the number of the measurements M should be anyway in the order of the degrees of freedom of the scattered field [7], which in the case of the slab can be calculated as [10]:

$$
I = 2L_{slab}\sqrt{\epsilon_b}\left(\frac{\lambda_M}{c_0} - \frac{\lambda_m}{c_0}\right), \tag{17}
$$

L_{slab} being the extent of the inhomogeneous slab, and λ_m and λ_M the minimum and maximum wavelength used to probe the scenario. Accordingly, it is possible to solve the inverse linear problem

$\Phi \Psi s = \mathcal{P}s = f$ by means of different optimization constrained problems [23,24] based on the minimization of the ℓ_1-norm of the unknown when represented in a proper sparse basis. The commonly adopted one in imaging problems, is the rectangular pulse basis function, commonly known as pixel and voxel in 2D and 3D, respectively. However, adopting such a basis, sparse recovery approaches are successful only if the scatterer resembles a point-like scatterer, whereas, in the case at hand, the plasma embeds almost the whole chamber which, in turn, is several probing wavelengths extended.

On the other hand, if a step-wise constant representation of the unknown profile is adopted, which is a reasonable assumption for the problem at hand, ℓ_1 optimization can be still exploited. One possibility is offered CS-based approaches involving as objective function the first derivative of the unknown [7]. For these reasons, we could consider the following recovery problems:

$$\text{argmin} \left\| \frac{d(\tilde{\omega}_p^2)}{dz} \right\|_{\ell_1} \qquad \text{subject to} \qquad \|\mathcal{P}\tilde{\omega}_p^2 - f\|_2 \leq \delta, \quad \tilde{\omega}_p^2 \geq 0, \qquad (18)$$

wherein $\tilde{\omega}_p^2 = \frac{\omega_p^2}{\omega_{max}^2}$. In (18), $\| \cdot \|_{\ell_1}$ and $\| \cdot \|_2$ denote the ℓ_1- and L^2-norms, respectively, and the parameter δ, which depends on the measurement error (noise on data) and model error (field approximation), is a positive defined parameter. In (18) the minimization of the ℓ_1-norm promotes the search of solutions with a sparse derivative among all the solutions consistent with the measured data (within the given error threshold δ). In addition, taking into account the real positive value of the unknowns, a further constraint ($\tilde{\omega}_p^2 \geq 0$) can be enforced in (18) in order to reduce the search space by convex optimization and improve the reconstruction results. The approach (18) is known as basis pursuit denoising (BPDN) or least absolute shrinkage and selection operator (LASSO) problem [24].

In [7], the sparsity-promoting approaches (18) have been proved to be capable in retrieving synthetic step-wise constant axial plasma profile which can be represented with few coefficients in terms of their first derivative. However, as the 1D single view multifrequency inverse scattering problem is severe limited, by its very nature, in terms of number of degrees of freedom (I), the number of non-null coefficient of any suitable representation basis should stay very few in order to achieve reliable reconstructions. This means that when smoother ad smoother profiles are considered, the adopted sparsity-promoting approach (18) is expected to be unsatisfactory.

For this reason, a "relaxed" version of (18) can be considered, i.e.,:

$$\text{argmin} \, \|\mathcal{P}\tilde{\omega}_p^2 - f\|_2 \qquad \text{subject to} \qquad \left\| \frac{d(\tilde{\omega}_p^2)}{dz} \right\|_{\ell_1} \leq \gamma, \quad \tilde{\omega}_p^2 \geq 0, \qquad (19)$$

wherein the role of the objective function with that of the constraints are exchanged. Approach (19) is expected to work better than (18) with profiles which are not exactly step-wise constant. Also, in this case, the assessment of the parameter γ is not a simple task. In this respect, prior information can be exploited to assess the optimal choice of the regularization parameter that mainly depends on the sparsity of the unknown at hand. However, the optimal choice of this threshold cannot be a priori established, pointing out one could a priori establish the range of the threshold and compare the results obtained for different values of γ.

4. Numerical Assessment Towards Benchmark Examples

The two benchmarks we refer to as "flat-top" and "hollow-core", respectively, represent expected electron density axial distribution in ECRIS-like devices [5]. The synthetic data have been generated considering the convolution between a rectangular window and a Hann window large 10 cm and 0.7 cm, respectively with a discretization grid of 256 cells. As a result, smoother profiles than those tackled in [22] can be considered. The total and the background fields have been simulated in the frequency range [17.5–34.5] GHz considering 87 evenly spaced frequency values in the solution of the forward problem stated by (3), thus the scattered field data has been obtained as difference between them. The solution of the forward problem can be performed by means of a standard conjugate gradient scheme as that proposed in [16]. The choice of the adopted frequency band is related to several requirements. The first one is the need to consider magnetically confined plasma as isotropic and weak scattering medium, and such assumption is fulfilled when the probing frequency is much higher than the cyclotron frequency, hereafter considered at 2.45 GHz. The second one, is related to the need for working with a large bandwidth in order to achieve the number of degrees of freedom I as large as possible, see Equation (17). Last, but not least, at this frequency band, also in centimeter-sized compact reactors, the propagation can be reasonably approximated as one-dimensional one along the chamber's axis. According to the above reasonings, we have set the frequency band of the fundamental mode in a rectangular waveguide WR34 that can be used to feed a horn antennas opening in the front and back walls of the plasma reactor [3].

We have considered three kind of measurement/background configurations which resemble possible experimental setup:

1. A single transmitting and receiving antenna measuring the reflection coefficient in a free-space homogeneous background (reflection-only measurement);
2. Single transmitting and receiving antenna measuring the reflection coefficient in presence of a PEC surface;
3. A transmitting and receiving antenna measuring the reflection coefficient and a receiving antenna measuring the transmission coefficient in a homogeneous free-space background (reflection and transmission measurement).

The three cases above can be seen a possible operational scenario where the reflection from metallic environments introduced by the plasma reactor can be neglected or not. This is the case, for example, when large and small reactors are in order, respectively.

To simulate the above measurement/background configurations, an absorbing, or free-space condition, can be very easily introduced, in the case at hand, by considering losses in PML layers ($z_1 - d \leq z \leq z_1$ and $z_2 \leq z \leq z_2 + d$) where the medium is gradually modified introducing losses according appropriate permittivity and permeability $\epsilon_r[1 - j\sigma(z)]$, $\mu_r[1 - j\sigma(z)]$ to have ideally no reflections from the PML [11,12,14].

For the sake of simplicity, we have chosen a simple polynomial grading for the PML layers conductivity profiles: $\sigma(z) = \sigma_{max}\left(\frac{z_1 - z}{d}\right)^m$ for $z_1 - d \leq z \leq z_1$ and $\sigma(z) = \sigma_{max}\left(\frac{z - z_2}{d}\right)^m$ for $z_2 \leq z \leq z_2 + d$; but more complex profiles can be used [25]. On the other hand, when PEC layer is considered, the PML is completely removed and perfect reflecting surface boundary conditions are enforced at $z = z_2$.

It is worth pointing out that for the solution of the forward problem, the plasma profiles are assumed to be lossy with the collision rate profile one order of magnitude smaller than the plasma angular frequency. As a result, this introduces a model error in the scattered field data which avoids the so-called "inverse crime". Furthermore, the scattered field data have been corrupted by random gaussian white noise of 5%. Finally, the number of the cells in the computational grid are set to $N = 256$.

Problem (19) has been solved by means of the environment CVXPY 1.0 for the solution of convex optimization problems [26]. In doing this, the choice of the threshold value γ has been set equal to the exact ℓ_1-norm of the actual unknown. Although this exact value cannot be chosen in any actual experiment, it allows an understanding of how the value of γ can be set on the base of expected a priori information on the degree of sparsity of the profile. Furthermore, for our investigation, such a choice allows the making of a fair comparison among the achieved results in the three different cases. For the two benchmarks profiles at hand it is $\gamma = 6.22$ (flat-top) and $\gamma = 7.14$ (hollow-core)

Finally, the required time to solve the inverse scattering problems takes very negligible time (below few seconds on a standard PC) due to the one-dimensional nature of the problem and then the size of the relevant matrix operator.

4.1. Reflection-Only Measurement with Single Antenna

We have considered free-space background with PML layers with thickness of 2 cm, $\sigma = 6$, $m = 4$, for a sketch the reader can refer to Figure 1a without considering the rightmost antenna. On the other hand, we have considered a non-homogeneous background with a PEC boundary condition on the right hand side of the computational domain ($z_2 = 25$ cm) to simulate the plasma chamber with only one antenna, see Figure 1b.

The imaging domain, $[z_{min}, z_{max}]$, is long 20 cm with the source point $z^{(\mathrm{View})}$ and the measurement point $z^{(\mathrm{Meas})}$ placed at 2.5 cm far from the imaging domain both in the case of single antenna and two antennas.

The imaging results are reported in Figures 2 and 3 for both the homogeneous and non-homogeneous background, respectively. As it can be see, in the case of PEC layer, the reconstructions brings better result with a reconstruction error (defined as $err = \frac{\|x_{true} - x_{rec}\|_2^2}{\|x_{true}\|_2^2}$) of 9.39% for the flat-top and 13.03% for the hollow core, with the plasma profile not heavily underestimated as in the case of the homogeneous background, wherein the reconstruction errors achieve 47.85% for the flat-top and 27.02% for the hollow core. This means that the presence of a reflecting surface allows to the achievement of better results in terms of accuracy of the reconstructed profiles. On the other hand, it is worth underlining that while the reconstruction of the flat-top profile is almost satisfactory, Figure 3a, in the case of the hollow-core profile, neither the presence of the PEC allows imaging of the main feature, although differences about the two different recovered profiles are evident, see Figure 3a,b. This may be due to a lower degree of sparsity of the hollow-core profile than that of the flat-top one and to the requirement by (19) to find solutions (profiles) with bounded ℓ_1 norm compatible with the objective function.

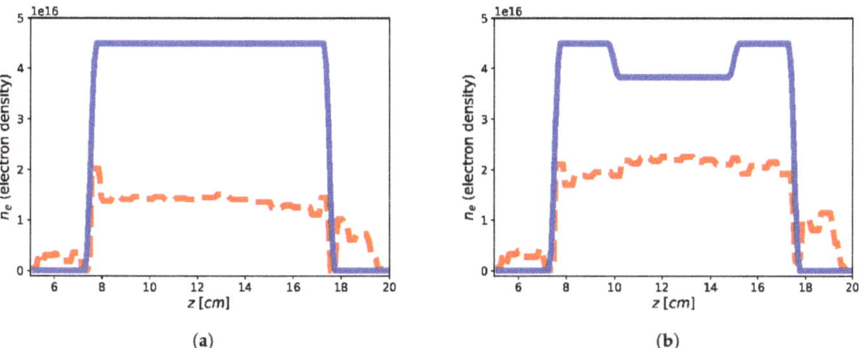

(a)   **(b)**

Figure 2. Reconstruction of the flat-top (**a**) and hollow core (**b**) plasma profiles using one tx-rx antenna in homogeneous vacuum background by means of approach (19). Blue continuous line: electronic density profile, red dashed line: retrieved electronic density profile.

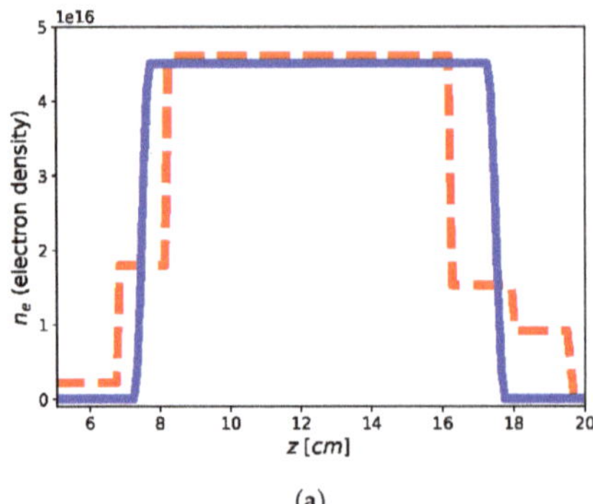 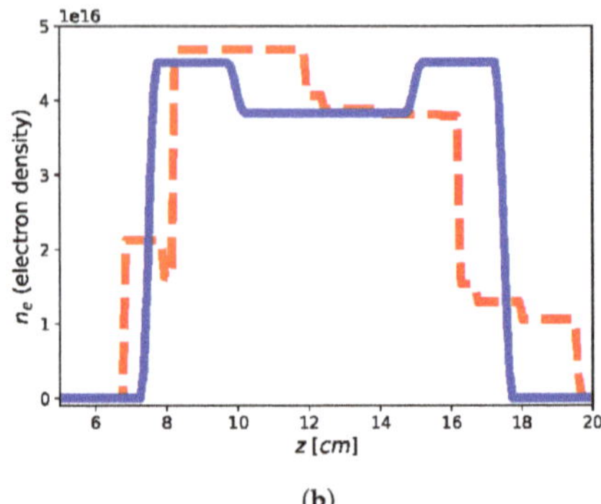

(a)   (b)

Figure 3. Reconstruction of the flat-top (**a**) and hollow core (**b**) plasma profiles using single tx-rx antenna in non-homogeneous background with a PEC reflecting surface by means of approach (19). Blue continuous line: electronic density profile, red dashed line: retrieved electronic density profile.

4.2. Reflection-Transmission Measurement with Two Antennas

In this case, two antennas are considered, see Figure 1a, with one (active) antenna radiating the probing field and measuring the reflection coefficient, and a second (passive) antenna measuring only the transmitting coefficient. Opposite to the case with a single antenna, in this case the non-homogeneous background accounting for two PEC surfaces has not been considered. First, from a modeling point of view, the use of PEC surfaces entails an electromagnetic cavity which enforces solutions only at the resonance frequencies given by $f_n = \frac{nc_0}{2L_{\text{cavity}}}$, i.e., when the cavity length (L_{cavity}) is an integer multiple of the probing half-wavelength, for a given finite set of integer n^* indexes depending on the adopted frequency band. Secondly, when the plasma partially embeds the cavity, the resonance frequencies shift in frequency, so that the expected large error on the background field approximation prevent using a linear model such as the Rytov approximation. Finally, from an applicative point of view, since in general, the front and back walls of the plasma reactors host two aperture antennas [3], just a (possible small) fraction of the transmitted microwave power will be reflected, while the most part will be gathered by the antennas. For these reasons, we have considered only homogeneous background.

The imaging results are reported in Figure 4 and the reconstruction errors achieve $err = 1.01\%$ and $err = 3.56\%$ for the flat-top and hollow core, respectively. Accordingly, it can be shown as the presence of a second passive antenna achieve better results in terms of accuracy of the retrieved profiles when compared to the single antenna in free-space homogeneous background. On the other hand, while the reconstruction of the flat-top profile is nearly optimal, also in this case, the reconstruction of the hollow-core profile is not satisfactory to apprise the main feature of the profile, although some small differences between the two retrieved profile are evident (as in the case of single antenna and PEC surface) and the same possible explanations hold true.

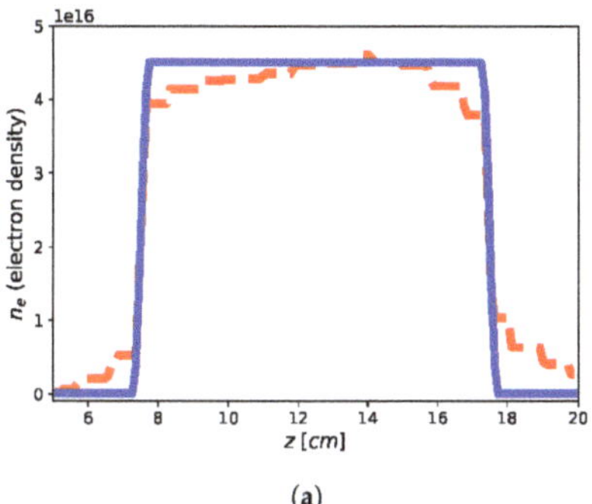 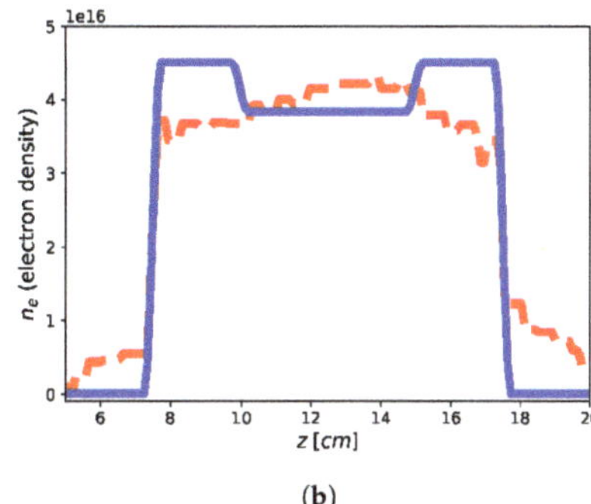

(a)   (b)

Figure 4. Reconstruction of the flat-top (**a**) and hollow core (**b**) plasma profiles using two antennas in homogeneous free-space background by means of approach 19. Blue continuous line: electronic density profile; red dashed line: retrieved electronic density profile.

5. Discussion of the Results through Singular Value Decomposition Analysis

To understand the different results achieved under the two different considered measurement configurations and related backgrounds, we have considered the SVD analysis of the relevant scattering operator $\mathcal{P}$ for the different analyzed cases.

First of all, it is worth underlining that even if we are not considering minimum energy solution approaches (L^2-norm), the SVD analysis can be retained still valid to understand the main features among the different measurement configurations and backgrounds, since we are considering sparse promoting approaches rather than exact compressive sensing-based approaches. Indeed, (19) entails a standard minimum energy solution objective function equipped with a regularization (penalty) ℓ_1-norm enforcing step-wise sparsity. For this reason, we numerically perform the SVD of the relevant scattering operator with the aim to give a qualitative interpretation of the different results shown in the previous Section.

Let us denote with $\{u_n, \sigma_n, v_n\}$ the singular value decomposition of the matrix operator $\mathcal{P}$ mapping the data-to-unknown relationship in (19), wherein u_n represents the left singular vectors (spanning the object space), v_n the right singular vectors (spanning the data space) and σ_n the singular values, in such way $\mathcal{P}u_n = \sigma_n v_n$ and $\mathcal{P}^+ v_n = \sigma_n u_n$, wherein $^+$ stand for the adjoint of the operator [18].

For all the measurement configurations adopted in the numerical analysis, we have analyzed the following metrics based on the SVD analysis, i.e.,:

1. *behavior of the singular values* —the logarithmic plot of the singular values as ordered in non-increasing fashion. Indeed, as the scattering operator is a compact one [18], its singular values exhibits an exponential decay after a given threshold index I (analytically expressed by (17)), which indicates the maximum number of the degrees of freedom, and hence of the parameters which can be conveyed back by the recovery procedure;

2. *spectral coverage* (SC) defined as:

$$SC = \sum_{n=0}^{N_T} |\tilde{U}_n|^2 \tag{20}$$

wherein $\tilde{U}$ is the Fourier Transform of the left singular vectors and N_T the truncation index used as regularization parameter in TSVD approach [10]. It is a measure of the class of profiles which can be actually retrieved in the object domain by the inversion process [10].

3. *point spread function* (PSF) defined as [27]:

$$PSF(z - z_0) = \sum_{n=0}^{N_T} u_n(z)u_n(z_0)^*$$

(21)

wherein $()^*$ stands for conjugation, and z_0 the abscissa with respect the PSF is considered. For the case at hand, we set $z_0 = 12.5$ cm, which is the center of the imaging domain. The point spread function is a direct measure of the ultimate attainable spatial resolution [27].

The SVD analysis is reported in Figure 5. It can be seen that the singular values are different for the considered configurations. Indeed, for the PEC case the lower order singular values show a greater magnitude while vanishing to zero (like in the case of the free-space background configuration). On the other hand, in the case of two antennas in homogeneous background, the magnitude of the singular values keeps almost always greater than the single antenna (without and with PEC).

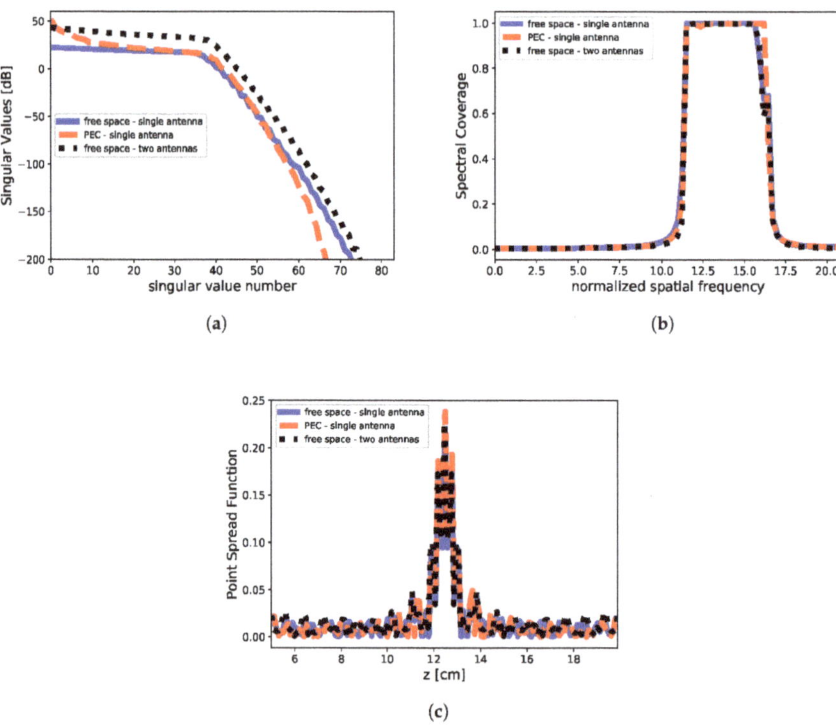

Figure 5. Singular value decomposition analysis. (**a**) Singular values, (**b**) spectral coverage and (**c**) point spread function in the case of single antenna free-space background (solid blue line), single antenna inhomogeneous PEC surface (red dashed line) and two antennas free-space background (dot black line). The cutoff value is $N_T = 40$ according to the change of slope in the singular values behavior, while the spatial frequency in SC are normalized to the extent of the imaging domain.

The SVD analysis give us the fundamental answer to understand the differences in the reconstruction process for the three adopted measurement configurations. Indeed, two main comments are in order. First, the presence of a perfect reflecting plane (PEC) introduces an increase in the singular values magnitude as compared to the homogeneous case. This is actually shown by the blue and red dashed line in Figure 5a. Interestingly, such an increase entails the advantage to deal with a more stable inverse scattering problem and with an increase of the information content [9]. Second, when we move to the case of two antennas, it can be easily understood as this configuration is almost equivalent to the case of single antenna operating in presence of a PEC surface. Indeed, this can be explained considering that the PEC surface acts as a "secondary" antenna which radiates (reflects) the field in such a way a virtual transmission coefficient can be measured by the actual antenna, as superimposed to the reflection coefficients that it would be measured without the PEC surface. As a result, when two antennas are considered, the reconstruction result achieves better results as in presence of a PEC surface. This analysis shows that the use of a second passive antenna is almost equivalent to have a reflecting surface behind the plasma slab. Anyhow, in all those cases wherein free-space approximation can be assumed (f.i., large meter-sized machine), two antennas can be used to achieve an enhancement in the reconstruction process. This has an immediate fallout in the design of the measurement setup depending on the constraints enforced by the experimental facilities.

On the other hand, the SVD analysis shows also that SC and PSF attained in three measurement configurations are identical, unless some very negligible differences due to numeric evaluation of the singular value system. These are confirmed by the same reconstruction performance achieved by the single antenna with PEC surface and two antennas in homogeneous background, for which the harmonic content is almost exactly the same. As a result, the improvement in the reconstruction in presence of a PEC surface can be attributed only to the difference in the singular values spectrum.

6. Conclusions

In this paper, we have addressed microwave imaging profilometry through the solution of an inverse scattering problem with a finite-difference formulation which takes into account, under the adopted assumptions, the effects of the metallic case represented by the plasma reactor. It has been shown that the presence of a metallic surface profoundly benefits the inversion recovery process when faced with sparsity-promoting approaches. In addition, the developed analysis shows the equivalence among two possible measurement configurations which are a single transmitting-receiving antenna in presence of a PEC surface and two antennas (one active and the second passive) able to measure also the transmission coefficients. Although the imaging capabilities are enhanced in presence of totally reflecting surface and with two antennas, microwave imaging profilometry still shows severe limitations in terms of accuracy in the recovery of arbitrary-shaped electron density profiles, i.e., not exactly step-wise constant profiles. Further activities are currently being devoted to exploit more efficient unknown representations and possible a priori information in order to achieve better reconstructions accuracy and to deal with a class of profiles wider than that considered in this paper.

Author Contributions: Conceptualization, L.D.D. and G.S.; methodology, L.D.D., D.M., A.F.M. and G.S.; writing—original draft preparation, L.D.D, A.F.M., D.M. and G.S. ; writing—review and editing, L.D.D. and G.S..

Funding: This research has been partially supported by the project CHANCE and PANDORA.

Acknowledgments: The authors would to thank L. Celona and G. Torrisi at the Laboratori Nazionali del Sud (Istituto Nazionale di Fisica Nucleare, Catania, Italy) for their fruitful suggestions and discussions.

Conflicts of Interest: The authors declare no conflict of interest.

References

1. Park, S.; Jang, J.; Choe, W. Sparse data recovery of tomographic diagnostics for ultra-large-area plasmas. *Plasma Sour. Sci. Technol.* **2019**, *28*, 035012. [CrossRef]
2. Geller, R. *Electron Cyclotron Resonance Ion Sources and ECR Plasmas*; Routledge: London, UK, 2018.
3. Torrisi, G.; Sorbello, G.; Leonardi, O.; Mascali, D.; Celona, L.; Gammino, S. A new launching scheme for ECR plasma based on two-waveguides-array. *Microw. Opt. Technol. Lett.* **2016**, *58*, 2629–2634. [CrossRef]
4. Mazzucato, E.; Munsat, T.; Park, H.; Deng, B.; Domier, C.; Luhmann, N., Jr.; Donné, A.; van de Pol, M. Fluctuation measurements in tokamaks with microwave imaging reflectometry. *Phys. Plasmas* **2002**, *9*, 1955–1961. [CrossRef]

5. Mascali, D.; Torrisi, G.; Leonardi, O.; Sorbello, G.; Castro, G.; Celona, L.; Miracoli, R.; Agnello, R.; Gammino, S. The first measurement of plasma density in an ECRIS-like device by means of a frequency-sweep microwave interferometer. *Rev. Sci. Instrum.* **2016**, *87*, 095109. [CrossRef] [PubMed]

6. Pastorino, M. *Microwave Imaging*; Wiley Online Library: Hoboken, NJ, USA, 2010.

7. Di Donato, L.; Morabito, A.F.; Torrisi, G.; Isernia, T.; Sorbello, G. Electromagnetic Inverse Profiling for Plasma Diagnostics via Sparse Recovery Approaches. *IEEE Trans. Plasma Sci.* **2019**, *47*, 1781–1787. [CrossRef]

8. Gilmore, C.; LoVetri, J. Enhancement of microwave tomography through the use of electrically conducting enclosures. *Inverse Probl.* **2008**, *24*, 035008. [CrossRef]

9. Solimene, R.; Maisto, M.A.; Pierri, R. Inverse scattering in the presence of a reflecting plane. *J. Opt.* **2015**, *18*, 025603. [CrossRef]

10. Pierri, R.; Persico, R.; Bernini, R. Information content of the Born field scattered by an embedded slab: multifrequency, multiview, and multifrequency—Multiview cases. *JOSA A* **1999**, *16*, 2392–2399. [CrossRef]

11. Sullivan, D.M. *Electromagnetic Simulation Using the FDTD Method*; John Wiley & Sons: Hoboken, NJ, USA , 2013.

12. Sacks, Z.S.; Kingsland, D.M.; Lee, R.; Lee, J.F. A perfectly matched anisotropic absorber for use as an absorbing boundary condition. *IEEE Trans. Antennas Propag.* **1995**, *43*, 1460–1463. [CrossRef]

13. Tonti, E. *The Mathematical Structure of Classical and Relativistic Physics*; Springer: Berlin, Germany, 2013.

14. Laudani, R.; Calvagna, A.; Sorbello, G.; Janner, D.; Tramontana, E. Analysis of the Discretization Error at Material Interfaces in Staggered Grids. *IEEE Trans. Antennas Propag.* **2010**, *58*, 1653–1661. [CrossRef]

15. Clemens, M.; Weiland, T. Discrete electromagnetics: Maxwell's equations tailored to numerical simulations. *Int. Compumag Soc. Newsl.* **2001**, *8*, 13–20.

16. Abubakar, A.; Hu, W.; Van Den Berg, P.; Habashy, T. A finite-difference contrast source inversion method. *Inverse Probl.* **2008**, *24*, 065004. [CrossRef]

17. Colton, D.; Kress, R. *Inverse Acoustic and Electromagnetic Scattering Theory*; Springer-Verlag: Berlin, Germany, 1992.

18. Bertero, M.; Boccacci, P. *Introduction to Inverse Problems in Imaging*; Institute of Physics: Bristol, UK, 1998.

19. Slaney, M.; Kak, A.C.; Larsen, L.E. Limitations of imaging with first-order diffraction tomography. *IEEE Trans. Microw. Theory Tech.* **1984**, *32*, 860 – 874. [CrossRef]

20. Stix, T. *Waves in Plasmas*; American Institute of Physics: College Park, MD, USA, 1992.

21. Di Donato, L.; Palmeri, R.; Sorbello, G.; Isernia, T.; Crocco, L. Assessing the capabilities of a new linear inversion method for quantitative microwave imaging. *Int. J. Antennas Propag.* **2015**, *2015*. doi:10.1155/2015/403760. [CrossRef]

22. Di Donato, L.; Palmeri, R.; Sorbello, G.; Isernia, T.; Crocco, L. A new linear distorted-wave inversion method for microwave imaging via virtual experiments. *IEEE Trans. Microw. Theory Tech.* **2016**, *64*, 2478–2488. [CrossRef]

23. Donoho, D.L. Compressed sensing. *IEEE Trans. Inf. Theory* **2006**, *52*, 1289–1306. [CrossRef]

24. Baraniuk, R.G. Compressive sensing [lecture notes]. *IEEE Signal Process. Mag.* **2007**, *24*, 118–121. [CrossRef]

25. Rumpf, R.C. Simple implementation of arbitrarily shaped total-field/scattered-field regions in finite-difference frequency-domain. *Prog. Electromagn. Res.* **2012**, *36*, 221–248. [CrossRef]

26. CVXPY 1.0. 2019. Available online: https://www.cvxpy.org/ (accessed on 9 July 2019).

27. Pierri, R.; Brancaccio, A.; Leone, G.; Soldovieri, F. Electromagnetic prospection via homogeneous and inhomogeneous plane waves: the case of an embedded slab. *AEU-Int. J. Electron. Commun.* **2002**, *56*, 11–18. [CrossRef]

Journal of ***Imaging***

Article

Developments in Electrical-Property Tomography Based on the Contrast-Source Inversion Method

Reijer Leijsen [1,†], **Patrick Fuchs** [2,†], **Wyger Brink** [1], **Andrew Webb** [1] and **Rob Remis** [2,*]

1   Department of Radiology, C.J. Gorter Center for High Field MRI, Leiden University Medical Center, 2333ZA Leiden, The Netherlands; R.L.Leijsen@lumc.nl (R.L.); W.M.Brink@lumc.nl (W.B.); A.Webb@lumc.nl (A.W.)
2   Circuits and Systems Group, Delft University of Technology, Faculty of Electrical Engineering, Mathematics and Computes Science, 2628CD Delft, The Netherlands; P.S.Fuchs@tudelft.nl
*   Correspondence: r.f.remis@tudelft.nl; Tel.: +31-15-278-1442
†   These authors contributed equally to this work.

Received: 4 December 2018; Accepted: 24 January 2019; Published: 1 February 2019

Abstract: The main objective of electrical-property tomography (EPT) is to retrieve dielectric tissue parameters from $\hat{B}_1^+$ data as measured by a magnetic-resonance (MR) scanner. This is a so-called hybrid inverse problem in which data are defined inside the reconstruction domain of interest. In this paper, we discuss recent and new developments in EPT based on the contrast-source inversion (CSI) method. After a short review of the basics of this method, two- and three-dimensional implementations of CSI–EPT are presented along with a very efficient variant of 2D CSI–EPT called first-order induced current EPT (foIC-EPT). Practical implementation issues that arise when applying the method to measured data are addressed as well, and the limitations of a two-dimensional approach are extensively discussed. Tissue-parameter reconstructions of an anatomically correct male head model illustrate the performance of two- and three-dimensional CSI–EPT. We show that 2D implementation only produces reliable reconstructions under very special circumstances, while accurate reconstructions can be obtained with 3D CSI–EPT.

Keywords: electromagnetic inverse scattering problems; magnetic resonance imaging; electrical-property tomography; nonlinear optimization; contrast-source inversion

---

1. Introduction

The conductivity and permittivity values of different tissue types are of great importance in a variety of medical applications. In magnetic-resonance (MR) safety [1] and hyperthermia-treatment planning [2], for example, conductivity tissue profiles are required to determine the specific absorption rate (SAR). Conductivity may also serve as a biomarker in oncology or in acute-stroke imaging [3]. Permittivity is important since it affects the spatial distribution of the transmitted electromagnetic field responsible for spin excitation.

Typically, tissue conductivity and permittivity values are measured ex vivo for a particular range of frequencies [4]. Other methods require elaborate hardware, such as electrical impedance tomography (EIT) [5] or microwave-imaging methods [6]. The objective of electrical-property tomography (EPT) is to retrieve these dielectric tissue values in vivo using an MR scanner and standard measurement protocols [3,7]. Specifically, with an MR scanner, the so-called $\hat{B}_1^+$-field, defined as $\hat{B}_1^+ = (\hat{B}_x + j\hat{B}_y)/2$, can be measured at a particular frequency of an operation called the Larmor frequency. This frequency is proportional to the magnitude of static background field B_0 via relation $f = \gamma B_0$, where $\gamma = 42.577$ MHz T^{-1} is the proton gyromagnetic ratio divided by 2π, leading to MR operating frequencies of 128 and 298 MHz for a 3 and 7 T scanner, respectively.

Reconstruction of dielectric tissue parameters is based on the measured $\hat{B}_1^+$-field, and what sets EPT apart from other more common inversion and imaging problems is that the measured $\hat{B}_1^+$-field has its support inside the reconstruction domain. The EPT reconstruction problem therefore belongs to the class of so-called hybrid inverse problems [8], and several EPT techniques have been proposed to reconstruct the conductivity and permittivity profiles based on these internal $\hat{B}_1^+$ data. Loosely speaking, these techniques can be divided into local differential-based approaches (see, e.g., References [9–12]) and global integral-based approaches (see e.g., References [13–18]). Combinations of local and global methods have been developed as well [19,20].

In this paper, we focus on a global integral-based EPT reconstruction method called contrast source inversion (CSI)–EPT, where a CSI approach [21–23] is taken to solve an EPT reconstruction problem. In particular, in CSI–EPT, the reconstruction problem is formulated as an optimization problem in which an objective function is iteratively minimized. This objective function consists of a term that measures the mismatch between modeled and measured data (data mismatch), and a term that measures the discrepancy in satisfying Maxwell's equations within the reconstruction domain using a global integral field representation (consistency mismatch). Including the second consistency term in the objective function is crucial to the performance of CSI, as shown in Reference [24].

Minimization of the objective function is carried out by iteratively updating a contrast function, which describes the dielectric constitution of the body part of interest, and a so-called contrast source, which is the product of the contrast function and the electric-field strength. Updating takes place by fixing one variable and updating the other. More precisely, the contrast function is first fixed, the contrast source is updated, and subsequently the contrast source is fixed and the contrast function is updated.

The CSI–EPT method was originally introduced in Reference [14], where it was shown that CSI–EPT is able to reconstruct strongly inhomogeneous conductivity and permittivity profiles within the center slice of an object placed in the center of a body coil in a 3 T MR scanner. The method was initially implemented for E-polarized electromagnetic fields in two-dimensional (2D) configurations in which the electrical field is parallel to the bore axis (z-axis) and the magnetic field is purely transverse, because it is significantly less complex than full three-dimensional implementation. The use of a 2D approach was justified since it was shown that the electromagnetic field in the midplane of a birdcage coil essentially has an E-polarized field structure [25]. An efficient alternative to CSI–EPT, first-order induced-current EPT or foIC-EPT, was presented in Reference [20] as well. This method exploits the structure of the two-dimensional E-polarized field to efficiently reconstruct the tissue profiles in the midplane of the transmit coil. The foIC-EPT method is significantly faster than CSI–EPT and produces reconstructions in real time with essentially the same quality as 2D CSI–EPT.

The CSI–EPT method has recently been extended to three-dimensional (3D) configurations in Reference [26]. With this 3D implementation of CSI–EPT, volumetric conductivity and permittivity profiles are obtained, and it is no longer necessary to restrict the reconstruction domain to the midplane of a transmit coil. Moreover, 3D CSI–EPT is based on vectorial 3D Maxwell equations and no (E-polarized) field structure is assumed to be present as in the case of a 2D approach. Unfortunately, computation times dramatically increase compared with 2D CSI–EPT and foIC-EPT and, depending on the configuration, it may take 3D CSI–EPT hours or even days to converge even on dedicated high-performance computers or servers. Apart from possible preconditioning techniques that may be applied to accelerate the convergence of 3D CSI–EPT, 2D CSI–EPT or foIC-EPT may be preferable in practice, since reconstruction times are significantly shorter compared with 3D approaches.

In this paper, we thoroughly investigate this issue and compare reconstructions obtained with 2D CSI–EPT, foIC-EPT, and 3D CSI–EPT. Reconstruction artifacts in the conductivity and permittivity profiles, the modeled $\hat{B}_1^+$-field, and the internal electric field are carefully studied. Our analysis shows that only under very special conditions is a 2D approach justified. Even if the electromagnetic field has an E-polarized field structure in the midplane of the transmit coil, imposing a two-dimensional field structure is generally too limiting an approximation unless the body part of interest and transmit coil strictly satisfy the longitudinal invariance condition.

This paper is organized as follows. In Section 2, the 2D and 3D CSI–EPT method is briefly reviewed, and the governing integral representations are presented. A variant of 2D CSI–EPT, foIC-EPT, is also presented, and a detailed analysis of the performance of all three reconstruction methods is presented in Section 3 using a realistic head model from Virtual Family [27]. A discussion with conclusions can be found in Section 4. Finally, we note that the position vectors in 2D and 3D are denoted by ρ and $\mathbf{x}$, respectively, and we use an $\exp(+j\omega t)$ time convention.

2. Theory

As mentioned above, the CSI–EPT algorithm operates on two unknowns and is based on two fundamental equations. Specifically, the unknowns in CSI–EPT are contrast function $\hat{\chi}$ and the contrast source $\hat{w}$, and the fundamental equations are the data equation and object or state equation.

The contrast function describes the dielectric contrast of the body with respect to free space and is given by $\hat{\chi}(\mathbf{x}) = \varepsilon_r(\mathbf{x}) - 1 - j\sigma(\mathbf{x})/\omega\varepsilon_0$, where $\varepsilon_r(\mathbf{x})$ and $\sigma(\mathbf{x})$ are, respectively, the unknown relative permittivity and conductivity profiles of the body, ε_0 is the permittivity of free space, and ω is the Larmor frequency of operation. The contrast function has bounded domain $\mathbb{D}_{\mathrm{body}}$ that is occupied by the body as its support, that is, the contrast function vanishes for $\mathbf{x} \notin \mathbb{D}_{\mathrm{body}}$. Finally, we note that the contrast function is dimensionless, and that its real part is determined by the permittivity profile, while its imaginary part is determined by the body's conductivity profile.

The contrast source in CSI–EPT is defined as $\hat{w} = \hat{\chi}\hat{E}$, where $\hat{E}(\mathbf{x})$ is the electric-field strength. Note that it is common to refer to $\hat{w}$ as a contrast source even though it is expressed in volts per meter and is actually scaled electric-field strength. The electric-field strength is obviously also unknown, since the dielectric constitution of the body is unknown. Even though this field is not of primary interest in EPT, CSI–EPT does provide electric-field reconstructions that may be used to reconstruct the local time-averaged power density that is dissipated into heat [1].

To arrive at the two fundamental equations of CSI–EPT, we set up a scattering formalism in which we make use of the linearity of Maxwell's equations and exploit the fact that the body occupies a bounded domain $\mathbb{D}_{\mathrm{body}}$. In particular, we first determine the electromagnetic field that is present inside an empty birdcage coil. In practice, this so-called background field is computed using electromagnetic-simulation software and we denote it by $\{\hat{E}^b, \hat{B}^b\}$. We note that the assumption is made here that external currents are impressed and field-independent. Consequently, antenna loading is not directly taken into account. The total electromagnetic field in presence of the body is denoted by $\{\hat{E}, \hat{B}\}$, and using the linearity of Maxwell's equations this field can be written as

$$\{\hat{E}, \hat{B}\} = \{\hat{E}^b, \hat{B}^b\} + \{\hat{E}^{sc}, \hat{B}^{sc}\}, \tag{1}$$

where $\{\hat{E}^{sc}, \hat{B}^{sc}\}$ is the scattered electromagnetic field due to the presence of the body. For this field, we have the integral representations

$$\hat{B}^{sc}(\mathbf{x}) = \int_{\mathbf{x}' \in \mathbb{D}_{\mathrm{body}}} \underline{\hat{G}}^{BJ}(\mathbf{x}, \mathbf{x}') \cdot \hat{w}(\mathbf{x}') \, dV \quad \text{and} \quad \hat{E}^{sc}(\mathbf{x}) = \int_{\mathbf{x}' \in \mathbb{D}_{\mathrm{body}}} \underline{\hat{G}}^{EJ}(\mathbf{x}, \mathbf{x}') \cdot \hat{w}(\mathbf{x}') \, dV, \tag{2}$$

where $\underline{\hat{G}}^{EJ}$ and $\underline{\hat{G}}^{BJ}$ are essentially the electric current to the electric field and electric current to magnetic field Green's tensors of the background medium. Note that these are the Green's tensors of a homogeneous background medium, and the presence of the coil is not taken into account. Explicit expressions for these tensors are given below.

Having these integral representations at our disposal, we can now present the basic CSI–EPT equations. We start with the equation that relates the measured $\hat{B}_1^+$-field to the contrast source. In particular, using the integral representation for the scattered magnetic field of Equation (2), we have

$$\hat{B}_1^{+;\text{sc}}(\mathbf{x}) = \frac{\hat{B}_x^{\text{sc}} + j\hat{B}_y^{\text{sc}}}{2} = \frac{1}{2} \int_{\mathbf{x}' \in \mathbb{D}_{\text{body}}} \sum_{k=x,y,z} \left[\hat{G}_{xk}^{\text{BJ}}(\mathbf{x},\mathbf{x}') + j\hat{G}_{yk}^{\text{BJ}}(\mathbf{x},\mathbf{x}') \right] \hat{w}_k(\mathbf{x}') \, dV, \tag{3}$$

which can be written more compactly as

$$\hat{B}_1^{+;\text{sc}}(\mathbf{x}) = \mathcal{G}_{\text{data}}\{\hat{\mathbf{w}}\}(\mathbf{x}) \quad \text{for } \mathbf{x} \in \mathbb{D}_{\text{body}}, \tag{4}$$

where linear data operator $\mathcal{G}_{\text{data}}$ is implicitly defined in Equation (3). Equation (4) is known as the data equation and relates unknown contrast source $\hat{\mathbf{w}}$ to the scattered $\hat{B}_1^+$-field. Note that this scattered field is known, since $\hat{B}_1^{+;\text{sc}}(\mathbf{x}) = \hat{B}_1^+(\mathbf{x}) - \hat{B}_1^{+;b}(\mathbf{x})$ and the total $\hat{B}_1^+$-field is known through measurements, while background field $\hat{B}_1^{+;b}(\mathbf{x})$ is known through simulations. The real phase is generally not known in practice, and tranceive-phase approximation is often used, which can lead to reconstruction artefacts at higher frequencies [28].

The second basic CSI–EPT equation, called the object or state equation, is obtained from the integral representation for the scattered electric field as given by the second equation of Equation (2). Using the definition of the scattered electric field $\hat{\mathbf{E}}^{\text{sc}} = \hat{\mathbf{E}} - \hat{\mathbf{E}}^b$, this integral representation can be written as

$$\hat{\mathbf{E}}(\mathbf{x}) - \int_{\mathbf{x}' \in \mathbb{D}_{\text{body}}} \underline{\hat{\mathbf{G}}}^{\text{EJ}}(\mathbf{x},\mathbf{x}') \cdot \hat{\mathbf{w}}(\mathbf{x}') \, dV = \hat{\mathbf{E}}^b(\mathbf{x}) \tag{5}$$

and multiplying the above equation by contrast function $\hat{\chi}$, we arrive at

$$\hat{\mathbf{w}}(\mathbf{x}) - \hat{\chi}(\mathbf{x}) \int_{\mathbf{x}' \in \mathbb{D}_{\text{body}}} \underline{\hat{\mathbf{G}}}^{\text{EJ}}(\mathbf{x},\mathbf{x}') \cdot \hat{\mathbf{w}}(\mathbf{x}') \, dV = \hat{\chi}(\mathbf{x})\hat{\mathbf{E}}^b(\mathbf{x}) \quad \text{for } \mathbf{x} \in \mathbb{D}_{\text{body}}, \tag{6}$$

which can be written more compactly as

$$\hat{\mathbf{w}}(\mathbf{x}) - \hat{\chi}(\mathbf{x})\mathcal{G}_{\text{body}}\{\hat{\mathbf{w}}\} = \hat{\chi}(\mathbf{x})\hat{\mathbf{E}}^b(\mathbf{x}) \quad \text{for } \mathbf{x} \in \mathbb{D}_{\text{body}}, \tag{7}$$

where linear operator $\mathcal{G}_{\text{body}}$ is implicitly defined in Equation (6).

To summarize, the two fundamental unknowns in CSI–EPT are contrast function $\hat{\chi}$ and contrast source $\hat{\mathbf{w}}$, and the basic CSI–EPT equations are the data Equation (4) and the object Equation (7).

Now, suppose we have available an approximation for the contrast function and contrast source. We denote these approximants by $\tilde{\chi}$ and $\tilde{\mathbf{w}}$, respectively, and, in order to measure how well these approximations satisfy the data and object equations, we introduce the data and object residuals as

$$\hat{r}_{\text{d}}(\mathbf{x}) = \hat{B}_1^{+;\text{sc}}(\mathbf{x}) - \mathcal{G}_{\text{data}}\{\tilde{\mathbf{w}}\}(\mathbf{x}) \quad \text{for } \mathbf{x} \in \mathbb{D}_{\text{body}}, \tag{8}$$

and

$$\hat{r}_{\text{o}}(\mathbf{x}) = \tilde{\chi}(\mathbf{x})\hat{\mathbf{E}}^b(\mathbf{x}) - \tilde{\mathbf{w}}(\mathbf{x}) + \tilde{\chi}(\mathbf{x})\mathcal{G}_{\text{body}}\{\tilde{\mathbf{w}}\}(\mathbf{x}) \quad \text{for } \mathbf{x} \in \mathbb{D}_{\text{body}}, \tag{9}$$

respectively, and measure their magnitudes using L_2-norms

$$\|\hat{r}_{\text{d}}\|_{\text{body}}^2 = \int_{\mathbf{x} \in \mathbb{D}_{\text{body}}} |\hat{r}_{\text{d}}(\mathbf{x})|^2 \, dV \quad \text{and} \quad \|\hat{r}_{\text{o}}\|_{\text{body}}^2 = \int_{\mathbf{x} \in \mathbb{D}_{\text{body}}} |\hat{r}_{\text{o}}(\mathbf{x})|^2 \, dV. \tag{10}$$

In CSI–EPT, these norms are used to define objective function

$$F(\tilde{\chi}, \tilde{\mathbf{w}}) = \frac{\|\hat{r}_{\text{d}}\|_{\text{body}}^2}{\|\hat{B}_1^{+;\text{sc}}\|_{\text{body}}^2} + \frac{\|\hat{r}_{\text{o}}\|_{\text{body}}^2}{\|\tilde{\chi}\hat{\mathbf{E}}^b\|_{\text{body}}^2} \tag{11}$$

and the goal is to find a contrast function and contrast source that minimizes this objective function. We note that including the two-norm of the object residual in the objective function (second term on the right-hand side of Equation (11)) is crucial to the success of CSI, since it has been shown that a contrast-source inversion approach without this term produces unsatisfactory results in general [24].

In CSI–EPT, finding the desired contrast function is now realized by minimizing the objective function in a "fix-one-minimize-for-the-other" approach. The iterative process continues until a predefined maximum number of iterations or specified tolerance level of the objective function has been reached. Specifically, the basic CSI–EPT algorithm is as shown in Listing 1.

Listing 1. Contrast-source inversion–electrical-property tomography (CSI–EPT).

- Given initial guesses $\tilde{\chi}^{[0]}$ and $\tilde{\mathbf{w}}^{[0]}$ for the contrast function and contrast source, respectively,
- For $k = 1, 2, \dots$

  1. Fix the contrast to $\tilde{\chi}^{[k-1]}$ and update the contrast source according to the update formula

  $$\tilde{\mathbf{w}}^{[k]} = \tilde{\mathbf{w}}^{[k-1]} + \alpha^{[k]}\mathbf{v}^{[k]}.$$

  2. Compute the corresponding electric-field strength $\hat{\mathbf{E}}^{[k]}$ according to (cf. Equation (5))

  $$\hat{\mathbf{E}}^{[k]}(\mathbf{x}) = \hat{\mathbf{E}}^{b}(\mathbf{x}) + \mathcal{G}_{\text{body}}\{\tilde{\mathbf{w}}^{[k]}\}(\mathbf{x}).$$

  3. Knowing contrast source $\tilde{\mathbf{w}}^{[k]}$ and corresponding electric-field strength $\hat{\mathbf{E}}^{[k]}$, determine contrast function $\tilde{\chi}^{[k]}$ from constitutive relation $\tilde{\mathbf{w}}^{[k]} = \tilde{\chi}^{[k]}\hat{\mathbf{E}}^{[k]}$ by solving least-squares problem $\|\tilde{\chi}\hat{\mathbf{E}}^{[k]} - \tilde{\mathbf{w}}^{[k]}\|^2_{\text{body}}$ for minimum norm contrast function $\tilde{\chi}$.

  4. Stop if objective function is smaller than user-specified tolerance level, or if maximum number of iterations have been reached.

- End.

Polak–Ribière update directions are usually taken for update direction $\mathbf{v}^{[k]}$ in Step 1 of the algorithm, but Fletcher–Reeves or Hesteness–Stiefel update directions may be used as well. To determine these update directions, the gradient of $F(\tilde{\chi}^{[k-1]}, \tilde{\mathbf{w}})$ with respect to $\tilde{\mathbf{w}}$ at $\tilde{\mathbf{w}} = \tilde{\mathbf{w}}^{[k-1]}$ is required. Explicit expressions for this gradient and corresponding step length $\alpha^{[k]}$ can be found in Reference [23], for example.

Note also that, with Equation (5), the object residual can be written as $\hat{\mathbf{r}}_{\text{o}} = \tilde{\chi}\hat{\mathbf{E}} - \tilde{\mathbf{w}}$ and, in Steps 2 and 3, we find minimum-norm contrast function $\tilde{\chi}$ for which $\|\hat{\mathbf{r}}_{\text{o}}\|^2_{\text{body}}$ is minimized. This contrast function is generally sensitive to small perturbations in $\tilde{\mathbf{w}}$ at locations where the magnitude of the electric field strength is "small." To suppress this effect, we can alternatively update the contrast function at every iteration according to update formula

$$\tilde{\chi}^{[k]} = \tilde{\chi}^{[k-1]} + \beta^{[k]}u^{[k]}, \tag{12}$$

with $u^{[k]}$ the Polak–Ribière update direction for the contrast function, and $\beta^{[k]}$ its corresponding update coefficient. Such an approach usually has a regularizing effect and typically leads to smoother reconstructions.

2.1. Object and Data Operators in Three-Dimensional CSI–EPT

In three dimensions and with air as a background medium, the integral representations for the scattered fields, as given by Equation (2), take on form

$$\hat{\mathbf{B}}^{\mathrm{sc}}(\mathbf{x}) = \mathrm{j}\frac{\omega}{c_0^2}\boldsymbol{\nabla}\times\hat{\mathbf{A}}^{\mathrm{sc}}(\mathbf{x}) \quad\text{and}\quad \hat{\mathbf{E}}^{\mathrm{sc}}(\mathbf{x}) = (k_0^2 + \boldsymbol{\nabla}\boldsymbol{\nabla}\cdot)\hat{\mathbf{A}}^{\mathrm{sc}}(\mathbf{x}), \tag{13}$$

where c_0 is the electromagnetic-wave speed in vacuum, $k_0 = \omega/c_0$ the wave number in vacuum, and $\hat{\mathbf{A}}^{\mathrm{sc}}$ is the vector potential given by

$$\hat{\mathbf{A}}^{\mathrm{sc}}(\mathbf{x}) = \int_{\mathbf{x}'\in\mathbb{D}_{\mathrm{body}}}\hat{G}(\mathbf{x}-\mathbf{x}')\hat{\mathbf{w}}(\mathbf{x}')\,\mathrm{d}V, \tag{14}$$

with $\hat{G}$ the three-dimensional Green's function of the vacuum background domain given by

$$\hat{G}(\mathbf{x}) = \frac{\exp(-\mathrm{j}k_0|\mathbf{x}|)}{4\pi|\mathbf{x}|}. \tag{15}$$

Note that the nabla operators act on position vector $\mathbf{x}$ and not on integration variable $\mathbf{x}'$. Three-dimensional object operator $\mathcal{G}_{\mathrm{body}}$ can easily be identified from the second part in Equation (13). For data operator $\mathcal{G}_{\mathrm{data}}$, however, we have to substitute the x and y components of the scattered magnetic flux density in Equation (3) to obtain

$$\hat{B}_1^{+;\mathrm{sc}} = \frac{\omega}{c_0^2}\left(\partial^+ \hat{A}_z^{\mathrm{sc}} - \partial_z\hat{A}^{+;\mathrm{sc}}\right), \tag{16}$$

where $\partial^+ = \frac{1}{2}(\partial_x + \mathrm{j}\partial_y)$ and $\hat{A}^{+;\mathrm{sc}} = \frac{1}{2}(\hat{A}_x^{\mathrm{sc}} + \mathrm{j}\hat{A}_y^{+;\mathrm{sc}})$. From the above expression for the scattered $\hat{B}_1^+$-field, the 3D data operator $\mathcal{G}_{\mathrm{data}}$ can be identified. Note the particular structure of this operator: scattered $\hat{B}_1^+$-field originates from a difference between the transverse variations of the longitudinal vector potential $(\partial^+ \hat{A}_z^{\mathrm{sc}})$ and the longitudinal variations of the transverse vector potential $(\partial_z\hat{A}^{+;\mathrm{sc}})$.

2.2. Object and Data Operators in Two-Dimensional CSI–EPT

In various papers (see Reference [25], for example) it has been reported that the radio-frequency (RF) field in the midplane of a birdcage coil is essentially E-polarized, meaning that the electric-field strength only has a longitudinal component ($\hat{\mathbf{E}} = \hat{E}_z\mathbf{i}_z$), while magnetic -flux density only has x and y components ($\hat{\mathbf{B}} = \hat{B}_x\mathbf{i}_x + \hat{B}_y\mathbf{i}_y$). Additionally, in a two-dimensional configuration that is invariant in the z direction, external electric current densities with longitudinal components only generate E-polarized fields. Identifying the currents in the rungs of the birdcage coil with these z-directed external current sources and denoting the slice through the object that coincides with the midplane of the birdcage coil by $\mathbb{S}_{\mathrm{body}}$, it makes sense to assume that within this midplane the RF field is essentially two-dimensional and E-polarized with integral representations for the scattered fields given by

$$\hat{\mathbf{B}}^{\mathrm{sc}}(\boldsymbol{\rho}) = \mathrm{j}\frac{\omega}{c_0^2}\boldsymbol{\nabla}_{\mathrm{T}}\times\hat{\mathbf{A}}^{\mathrm{sc}}(\boldsymbol{\rho}), \quad\text{and}\quad \hat{\mathbf{E}}^{\mathrm{sc}}(\boldsymbol{\rho}) = k_0^2\hat{\mathbf{A}}^{\mathrm{sc}}(\boldsymbol{\rho}), \tag{17}$$

where $\boldsymbol{\rho}$ is the position vector in the midplane of the birdcage coil, $\boldsymbol{\nabla}_{\mathrm{T}} = \mathbf{i}_x\partial_x + \mathbf{i}_y\partial_y$ is the transverse nabla operator, and

$$\hat{\mathbf{A}}^{\mathrm{sc}}(\boldsymbol{\rho}) = \int_{\boldsymbol{\rho}'\in\mathbb{S}_{\mathrm{body}}}\hat{G}(\boldsymbol{\rho}-\boldsymbol{\rho}')\hat{\mathbf{w}}(\boldsymbol{\rho}')\,\mathrm{d}S \tag{18}$$

is the vector potential in two dimensions (and is thus expressed as a two-dimensional integral as opposed to the three-dimensional integral in the 3D case) with

$$\hat{G}(\boldsymbol{\rho}) = -\frac{\mathrm{j}}{4}H_0^{(2)}(k_0|\boldsymbol{\rho}|) \tag{19}$$

the Green's function of the two-dimensional homogeneous background medium (air) and $H_0^{(2)}$ is the Hankel function of the second kind and order zero. In this two-dimensional case, object operator $\mathcal{G}_{\mathrm{body}}$ can easily be identified from the second part of Equation (17) and does not contain a gradient-divergence operator as in the three-dimensional case. For 2D data operator $\mathcal{G}_{\text{data}}$, we have to substitute the x and y components of the magnetic-flux density as given by the first part of Equation (17) in the definition of the $\hat{B}_1^+$-field to obtain

$$\hat{B}_1^{+;\text{sc}} = \frac{\omega}{c_0^2}\partial^+ \hat{A}_z^{\text{sc}}. \tag{20}$$

From this expression, 2D data operator $\mathcal{G}_{\text{data}}$ can now easily be identified. Comparing the two-dimensional field representation of Equation (20) with its three-dimensional counterpart of Equation (16), we observe that longitudinal spatial variations are absent in the two-dimensional case. Moreover, the vector potentials in both expressions are different, since this quantity is computed using Equation (18) in the two-dimensional case, while the three-dimensional vector potential is given by Equation (14). Differences between two- and three-dimensional CSI–EPT reconstructions are discussed further in Section 3.

2.3. Simplified Two-Dimensional CSI–EPT—foIC-EPT

In two dimensions, the CSI–EPT algorithm can be simplified by exploiting the particular structure of E-polarized RF fields. To make this simplification explicit, we first introduce differentiation operator $\partial^- = \frac{1}{2}(\partial_x - j\partial_y)$ and note that operators ∂^- and ∂^+ essentially factor two-dimensional Laplacian $\Delta = \partial_x^2 + \partial_y^2$ as

$$\Delta = 4\partial^-\partial^+ = 4\partial^+\partial^-. \tag{21}$$

Now, as a first step, we substitute the second part of Equation (17) in Equation (20) to obtain

$$\hat{B}_1^{+;\text{sc}}(\boldsymbol{\rho}) = \frac{1}{\omega}\partial^+ \hat{E}_z^{\text{sc}}(\boldsymbol{\rho}). \tag{22}$$

Subsequently, we use the definition of the scattered fields to write the above expression as

$$\hat{B}_1^+(\boldsymbol{\rho}) = \hat{B}_1^{+;\text{b}}(\boldsymbol{\rho}) + \frac{1}{\omega}\partial^+ \hat{E}_z(\boldsymbol{\rho}) - \frac{1}{\omega}\partial^+ \hat{E}_z^{\text{b}}(\boldsymbol{\rho}) \tag{23}$$

and, since $\hat{B}_1^{+;\text{b}}(\boldsymbol{\rho}) = \frac{1}{\omega}\partial^+ \hat{E}_z^{\text{b}}(\boldsymbol{\rho})$, this simplifies to

$$\hat{B}_1^+(\boldsymbol{\rho}) = \frac{1}{\omega}\partial^+ \hat{E}_z(\boldsymbol{\rho}). \tag{24}$$

If we now act with the ∂^- operator on this equation, we obtain

$$\partial^- \hat{B}_1^+ = \frac{1}{4\omega}\Delta \hat{E}_z \tag{25}$$

and since $\hat{E}_z$ satisfies $\Delta \hat{E}_z - j\omega\mu_0 \hat{J}_z^{\text{ind}} = 0$ with

$$\hat{J}_z^{\text{ind}} = (\sigma + j\omega\varepsilon)\hat{E}_z, \tag{26}$$

we arrive at

$$\hat{J}_z^{\text{ind}} = \frac{4}{j\mu_0}\partial^- \hat{B}_1^+. \tag{27}$$

This last equation shows that, in two dimensions, induced current density is obtained (accounting for multiplication by $4/j\mu_0$) by acting with the ∂^- operator on the total B_1^+-field. The simplified CSI–EPT method is therefore called a first-order-induced current EPT method, since a first-order differentiation of the B_1^+-field essentially immediately results in an image of the induced current density.

As shown in Reference [20], after the induced current density is obtained, the corresponding electric-field strength can be computed by solving a specific integral equation defined on S_{body}. With the electric-field strength now known, the conductivity and permittivity profiles within the slice can be obtained from Equation (26). The overall first-order induced current density EPT algorithm can be summarized as presented in Listing 2.

Listing 2. First-Order Induced Current EPT Algorithm (foIC-EPT).

- Given the measured $\hat{B}_1^+$-field in the midplane of the birdcage coil:

  1. Determine the induced current density using Equation (27).

  2. Determine the corresponding electric-field strength by solving a specific integral equation (Equation (12) in Reference [20]).

  3. Knowing the induced current density and the electric-field strength, determine conductivity and permittivity profiles using Equation (26).

Further details about this algorithm can be found in Reference [20]. Finally, we note that the above algorithm is a direct noniterative EPT method and, as opposed to CSI–EPT, requires the solution of a system of equations (Step 2) to arrive at the reconstructed conductivity and permittivity profiles. Fortunately, as demonstrated in Reference [20], this system of equations can efficiently be solved using iterative solvers such as the generalized minimal residual (GMRES) method [29] and, typically, only a small number of iterations is required to reach a prescribed error.

3. Methods and Results

To illustrate the performance of foIC-EPT and two- and three-dimensional CSI–EPT, we reconstructed the conductivity and permittivity profiles of the head of anatomical human-body model Duke from Virtual Family [27] (see Figure 1a,b), from noisefree $\hat{B}_1^+$-data. The head model consists of $124 \times 100 \times 109$ isotropic voxels with side lengths of 2 mm. The model was placed inside an ideal high-pass birdcage coil (see Figure 1a) consisting of 16 rungs, each having a width of 25 mm. The coil has a radius of 150 mm, is 195 mm long, and is driven in quadrature at 128 MHz, which corresponds to the operating frequency of a 3 T MRI system. The shield surrounding the coil has a radius of 180 mm and length of 200 mm. Commercial EM simulation software (XFdtd, v.7.5, Remcom State College, PA, USA) was used to obtain the background field $\{\hat{E}^b, \hat{B}^b\}$ as generated by the high-pass birdcage coil. Finally, to investigate the difference between two- and three-dimensional conductivity and permittivity reconstructions, we also considered a longitudinally uniform "head model" in which the center slice was simply repeated in the longitudinal direction, thereby creating a model with no variations in the longitudinal z direction within the head (see Figure 1c).

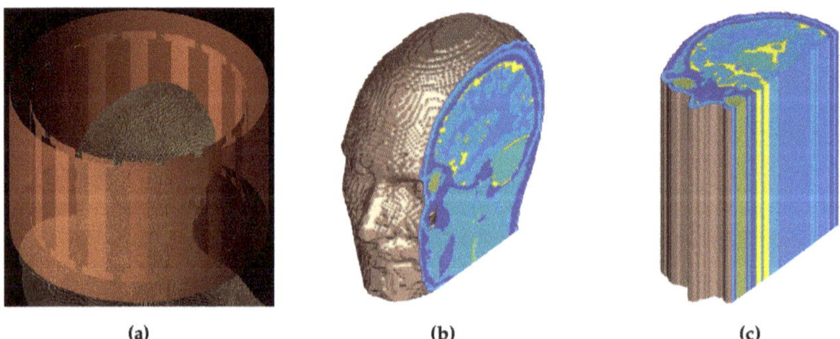

(a)   **(b)**   **(c)**

Figure 1. Birdcage coil and head models. (**a**) High-pass birdcage coil with the head model placed inside, (**b**) Duke head model from Virtual Family [27], and (**c**) longitudinally uniform head model obtained by repeating the center slice in the longitudinal direction.

3.1. Two-Dimensional CSI–EPT and foIC-EPT

The CSI–EPT method was originally implemented for two-dimensional configurations in Reference [14] to study its potential as an EPT reconstruction method and to test if the method can handle strongly inhomogeneous tissue profiles. Let us therefore start with a purely two-dimensional reconstruction problem in which we attempt to reconstruct conductivity and permittivity profiles within the center slice of the head model shown in Figure 2a. In this two-dimensional setting, we took the background field in the midplane of the realistic birdcage coil shown in Figure 1a as the 2D background field. The reconstructed conductivity and permittivity profiles obtained after 5000 iterations of the two-dimensional CSI–EPT method are shown in Figure 2b. It took the algorithm approximately 86 s on an Intel i7-6700 CPU (Intel, Santa Clara, CA, USA) operating on Windows 7 with Matlab 2016a (Mathworks, Natick, MA, USA) to arrive at these reconstructions, and we terminated the algorithm after 5000 iterations since the objective function had already dropped below a 1.53×10^{-5} tolerance level at that point and essentially no significant improvements were obtained. In addition, the foIC-EPT reconstruction profiles of conductivity and permittivity are shown in Figure 2c, and the errors of CSI–EPT and foIC-EPT conductivity and permittivity reconstructions are shown in Figure 2d,e, respectively. We observed that the quality of the foIC-EPT reconstructions was similar to CSI–EPT even though it took foIC-EPT only a fraction of a second to produce these reconstructions (see Table 1 for details).

Table 1. Mean and standard deviation of reconstructed electrical properties in different tissue that are apparent in the center slice of the head models using 2D CSI–EPT and foIC-EPT in a two-dimensional setting. Units of σ and ε_r are in siemens per meter and permittivity of free space, respectively.

	Conductivity (σ)			Relative Permittivity (ε_r)		
	True	**2D CSI–EPT**	**foIC-EPT**	**True**	**2D CSI–EPT**	**foIC-EPT**
Fat	0.07	0.13 ± 0.10	0.18 ± 0.10	12.37	17.23 ± 7.89	19.27 ± 7.15
Red marrow	0.16	0.11 ± 0.06	0.15 ± 0.03	13.54	11.08 ± 2.52	13.68 ± 1.47
Bone	0.07	0.14 ± 0.15	0.21 ± 0.14	14.72	19.09 ± 8.05	21.73 ± 6.90
Eye lens	0.31	0.54 ± 0.11	0.78 ± 0.09	42.79	48.86 ± 5.26	51.52 ± 1.44
Nerve	0.35	0.74 ± 0.30	0.77 ± 0.26	44.07	51.49 ± 8.74	47.33 ± 7.42
Connective tissue	0.50	0.47 ± 0.12	0.44 ± 0.13	51.86	46.48 ± 7.56	40.54 ± 7.06
White matter	0.34	0.36 ± 0.04	0.38 ± 0.04	52.53	54.33 ± 3.12	55.37 ± 3.54
Muscle	0.72	0.63 ± 0.11	0.55 ± 0.13	63.49	56.76 ± 7.28	48.30 ± 8.72
Eye sclera	0.92	0.89 ± 0.15	0.87 ± 0.13	65.00	56.78 ± 6.67	50.26 ± 5.95
Skin	0.52	0.48 ± 0.08	0.36 ± 0.09	65.44	59.00 ± 8.95	42.29 ± 8.55
Hypothalamus	0.80	0.88 ± 0.11	0.91 ± 0.11	66.78	59.48 ± 4.91	54.12 ± 4.20
Eye vitreous humor	1.51	1.46 ± 0.13	1.40 ± 0.15	69.06	66.19 ± 4.86	61.03 ± 4.95
Cornea	1.06	0.92 ± 0.13	0.83 ± 0.11	71.46	61.52 ± 9.32	52.81 ± 5.84
Gray matter	0.59	0.61 ± 0.15	0.63 ± 0.15	73.52	72.68 ± 4.18	70.54 ± 4.92
Midbrain	0.83	0.84 ± 0.17	0.88 ± 0.18	79.74	78.58 ± 10.24	81.93 ± 12.33
Cerebrospinal fluid	2.14	1.90 ± 0.29	1.75 ± 0.29	84.04	80.66 ± 8.57	76.46 ± 11.31
Mucosa	2.28	1.50 ± 0.03	1.01 ± 0.02	116.00	78.07 ± 5.17	53.83 ± 2.77

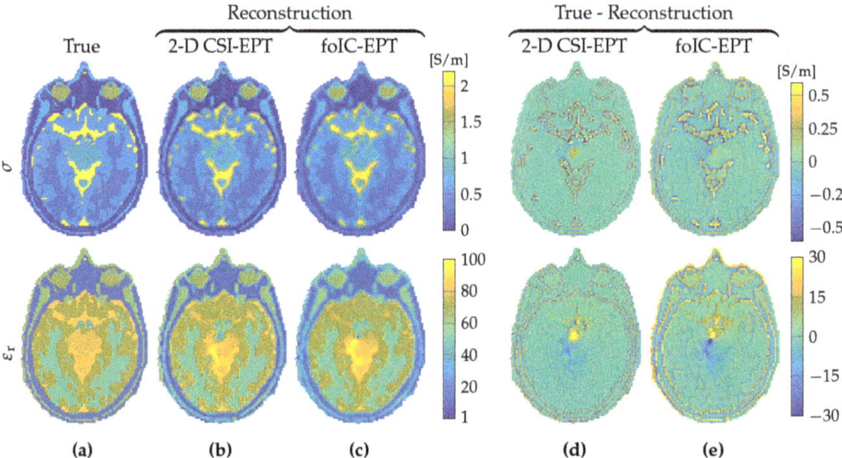

Figure 2. Reconstruction results from 2D reconstruction methods. (**a**) True model, (**b**) reconstruction obtained after 5000 iterations of 2D CSI–EPT, and (**c**) reconstruction from foIC-EPT. Respective errors are shown in (**d**,**e**). Top row shows conductivity and bottom row shows relative permittivity.

3.2. Three-Dimensional CSI–EPT

In a two-dimensional approach, the RF field is E-polarized with electric-field strength that is longitudinal ($\hat{\mathbf{E}} = \hat{E}_z \mathbf{i}_z$) and magnetic-flux density that is transverse ($\hat{\mathbf{B}} = \hat{B}_x \mathbf{i}_x + \hat{B}_y \mathbf{i}_y$). Such an approach was shown to be reasonable for a homogeneous cylindrical phantom in a central region of a body coil consisting of elementary center-fed dipole antennas in Reference [25]. Indeed, when the longitudinally uniform head model of Figure 1c was placed within our birdcage coil we also observed that the x and y components of the electric-field strength in the central transverse slice were small compared to its z component, as illustrated in the top rows of Figure 3a–c and Figure 4a,b.

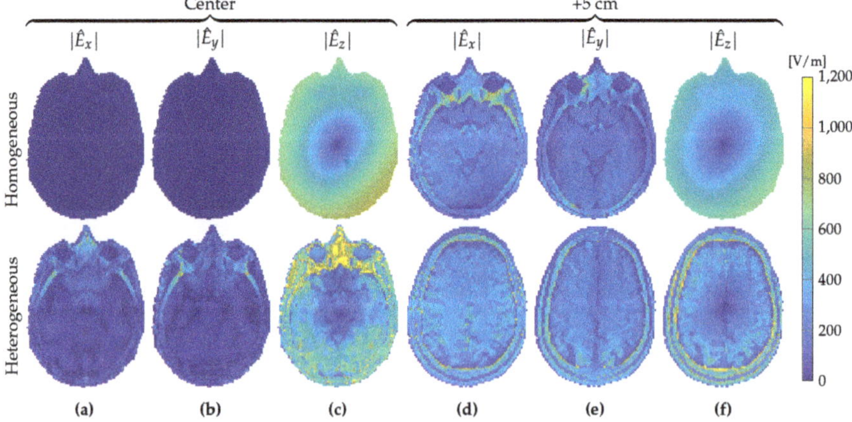

Figure 3. Magnitude of electric-field-strength components. x, y, and z components at (**a**–**c**) the transversal midplane, and (**d**–**f**) the slice 5 cm higher, respectively. Top and bottom row show fields in the case of a longitudinal homogeneous and heterogeneous object, respectively.

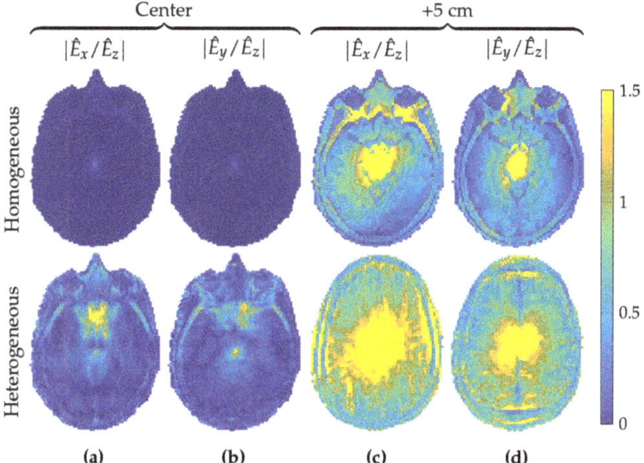

Figure 4. Ratios of x and y components of the electric-field strength relative to its z component. Relative field components at (**a**,**b**) the center slice and (**c**,**d**) the slice 5 cm higher. Top row is for the longitudinally uniform object, bottom row for the object with longitudinal variations.

However, as we move away from the center slice in the longitudinally uniform head model of Figure 1c, the magnitude of the x and y components of the electric-field strength starts to increase, as illustrated in the top rows of Figure 3d–f and Figure 4c,d, where the magnitude of the electric-field-strength components is shown in a slice located 5 cm above the central slice. We observed that even though the transverse components of the electric-field strength were negligible within the center slice, they could no longer be neglected when 5 cm away from it.

Furthermore, for the realistic heterogeneous head model of Figure 1b, a two-dimensional E-polarized field assumption completely failed, as shown in the bottom rows of Figures 3a–f and 4a–d. In the slice 5 cm above the central slice, and even within the central slice itself, the x and y components of the electric-field strength could no longer be neglected and had to be taken into account in the full Maxwell equations to properly describe RF field behavior within the head model.

To study the effects of longitudinal spatial variations of the tissue parameters on the $\hat{B}_1^+$-field, we considered Equation (16) again and write it in form

$$\hat{B}_1^{+;\mathrm{sc}} = \mathcal{B}^{\mathrm{tra}} + \mathcal{B}^{\mathrm{lon}}, \tag{28}$$

where $\mathcal{B}^{\mathrm{tra}} = \frac{\omega}{c_0^2}\partial^+ \hat{A}_z^{\mathrm{sc}}$ and $\mathcal{B}^{\mathrm{lon}} = -\frac{\omega}{c_0^2}\partial_z \hat{A}^{+;\mathrm{sc}}$. Longitudinal variation term $\mathcal{B}^{\mathrm{lon}}$ is absent in a 2D approach (see Equation (20)), since, in a 2D setting, the configuration is assumed to be invariant in the longitudinal z-direction ($\partial_z = 0$). Figure 5, however, shows that, for both the longitudinally homogeneous and realistic heterogeneous head model, the longitudinal variation term is significant and cannot be ignored. Especially near the periphery of both head models, $\mathcal{B}^{\mathrm{lon}}$ contributes to the scattered $\hat{B}_1^+$-field. More specifically, within a 1 cm outer boundary layer located in the center slice, the mean of fraction $|\mathcal{B}^{\mathrm{lon}}/\hat{B}_1^{+;\mathrm{sc}}|$ is 1.18 and 1.25 for the homogeneous and inhomogeneous head model, respectively; in the inner region, these means are 0.51 and 0.60, and similar averages were obtained for the slice located 5 cm above the center slice. From these observations, it is clear that longitudinal variations of transverse vector potential $\hat{A}^{+;\mathrm{sc}}$ contribute to the scattered $\hat{B}_1^+$-field and cannot be ignored.

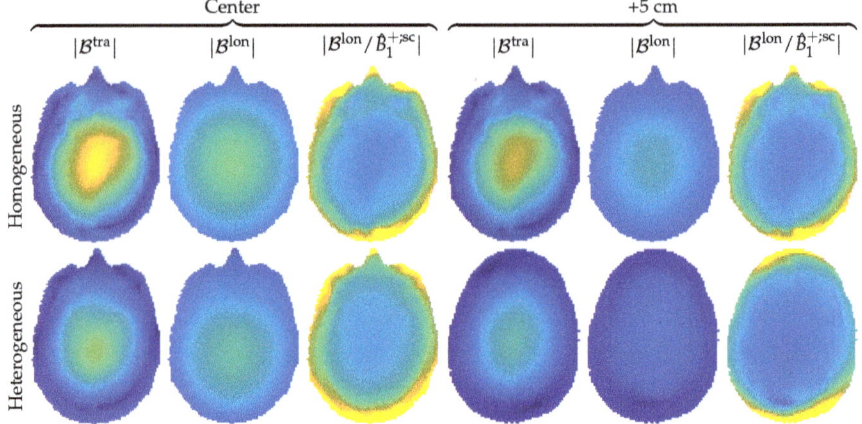
Center
+5 cm
$|\mathcal{B}^{\mathrm{tra}}|$
$|\mathcal{B}^{\mathrm{lon}}|$
$|\mathcal{B}^{\mathrm{lon}}/\hat{B}_1^{+;\mathrm{sc}}|$
$|\mathcal{B}^{\mathrm{tra}}|$
$|\mathcal{B}^{\mathrm{lon}}|$
$|\mathcal{B}^{\mathrm{lon}}/\hat{B}_1^{+;\mathrm{sc}}|$
Homogeneous
Heterogeneous

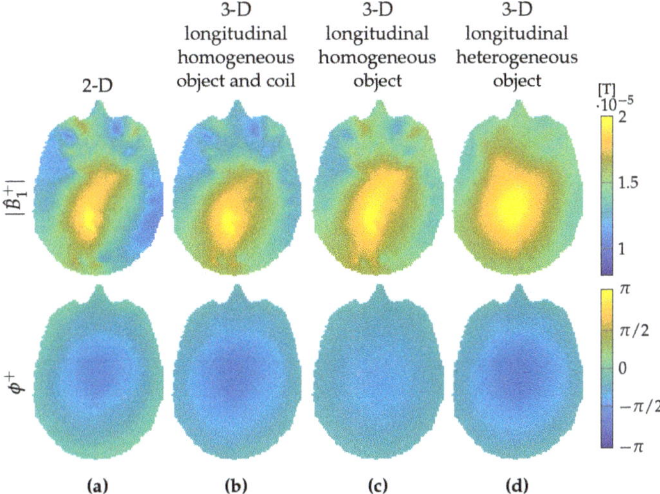

Figure 6. $\hat{B}_1^+$ field comparison. (**a**) Total $\hat{B}_1^+$ field assumed in the 2D setting, (**b**) total $\hat{B}_1^+$ field obtained in a 3D setting with longitudinal homogeneity of the object and coil, (**c**) of longitudinal homogeneity of only the object, and (**d**) with longitudinal variations also of the object. Top row shows the $\hat{B}_1^+$ magnitude, bottom row the $\hat{B}_1^+$ phase.

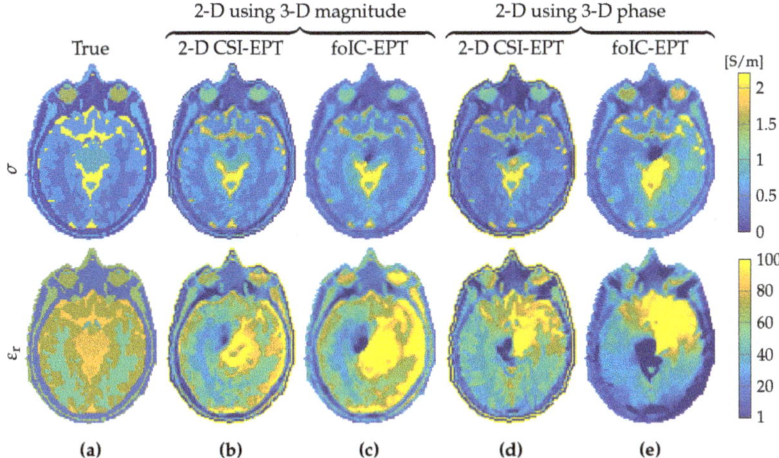

Figure 7. Reconstruction results from 2D reconstruction methods using parts of 3D $\hat{B}_1^+$ data. (**a**) True model, (**b,c**) reconstruction results assuming 2D phase with 3D magnitude, and (**d,e**) reconstruction results assuming 2D magnitude with 3D phase of the $\hat{B}_1^+$-field in the central transverse slice from simulations with the longitudinal invariant head model. Top row shows conductivity and the bottom row shows relative permittivity.

The reconstructions of the conductivity and relative permittivity profiles for the full 3D case without any further assumptions, using 3D magnitude as well 3D phase $\hat{B}_1^+$ data are shown for the longitudinally uniform model in Figure 8a–f and for the realistic heterogeneous head model in Figure 8g–l. Reconstructions are shown for the central slice profiles as well as for the profiles located within the slice positioned 5 cm above the central slice. For comparison, 2D CSI–EPT reconstructions based on 3D $\hat{B}_1^+$ data are also presented. Relative residual error (norm of the difference between the exact and reconstructed profile normalized by the norm of the exact profile, where the norm is taken over the center slice) of Figure 8h is 0.7339 and 0.8263 for the conductivity and permittivity, respectively, while the relative residual error of the conductivity and permittivity of Figure 8i is 0.3358 and 0.1587, respectively. Clearly, 2D CSI–EPT is unable to accurately reconstruct conductivity and permittivity profiles. The 2D and 3D permittivity reconstructions are also less accurate than conductivity reconstructions, indicating that $\hat{B}_1^+$-field data acquired at 3 T are less sensitive to permittivity variations.

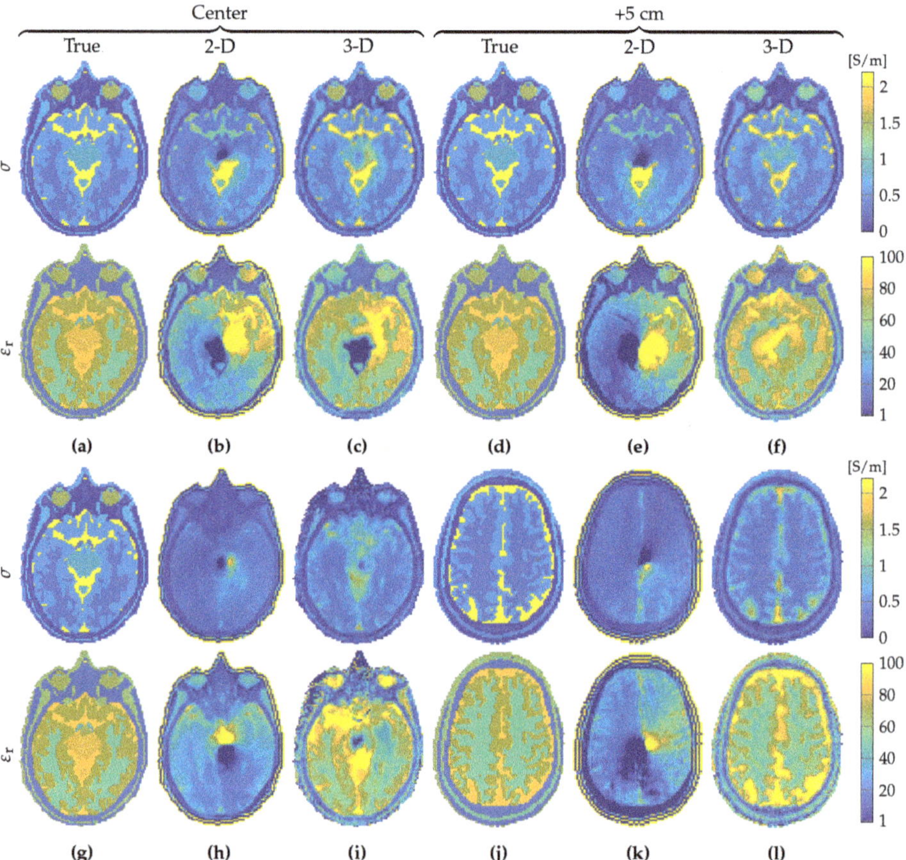

Figure 8. Reconstruction comparison of 2D and 3D CSI–EPT on 3D $\hat{B}_1^+$ fields after 5000 and 50,000 iterations, respectively. (**a–c**) true object, reconstruction with 2D CSI–EPT, and reconstruction with 3D CSI–EPT for a homogeneous object and for the center slice. (**d–f**) Same as above, but for a slice 5 cm higher. (**g–l**) same as (**a–f**), but in the case of a longitudinal inhomogeneous object. Top row depicts the conductivity, bottom row shows permittivity.

Finally, to emphasize that 3D CSI–EPT is a fully three-dimensional volumetric reconstruction method, we present a full 3D CSI–EPT reconstruction of the realistic head model obtained after 50,000 iterations based on 3D $\hat{B}_1^+$ data in Figure 9. This number of iterations was chosen due to time constraints, since it takes approximately 110 h on an Intel i7-6700 CPU operating on Windows 7 with Matlab 2016a.

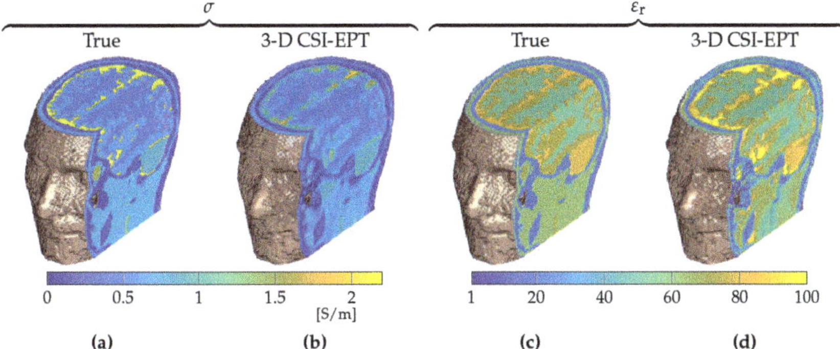

Figure 9. Three-dimensional visualization of a section of the 3D CSI–EPT reconstruction of the heterogeneous Duke head model (Figure 1b) after 50,000 iterations. (**a,b**) True and reconstructed conductivity, and (**c,d**) true and reconstructed relative permittivity. Top slice that is visible is the slice 5 cm above the transverse midplane.

4. Discussion

We investigated the performance of two- and three-dimensional CSI–EPT in reconstructing dielectric tissue profiles based on $\hat{B}_1^+$ data collected inside the reconstruction slice or domain of interest. Since these data has their support inside the reconstruction domain, EPT belongs to the class of so-called hybrid inverse problems [8]. In CSI–EPT, reconstructing tissue parameters is posed as an optimization problem in which an internal objective function, that is, an objective function that measures both field and model discrepancies within the domain of interest, is minimized in an iterative manner. Field discrepancies are measured by considering the L_2-norm of the difference between modeled and measured data, while model discrepancies are measured by an L_2-norm that tells us how well a conductivity and permittivity tissue profile, and the corresponding contrast source, satisfy Maxwell's equations. Including model discrepancies in the objective function is crucial to the performance of CSI–EPT, since it has been shown that, without this term, unsatisfactory reconstruction results may be obtained [24]. In addition to the tissue profiles, CSI–EPT reconstructs electric-field strength as well, and may therefore also be used to predict the SAR that is induced inside the body or a body part of interest [30], which is important for MR safety and hyperthermia-treatment planning, for example. Finally, we also showed that, in two dimensions, an alternative noniterative and integral-based reconstruction algorithm called foIC-EPT may be employed. This method is significantly faster than 2D and 3D CSI–EPT, and reconstructs tissue profiles and corresponding electric-field strength essentially in real time on a present-day standard laptop or PC (Intel i5-i7 or similar). However, foIC-EPT is restricted to two-dimensional configurations since it exploits two-dimensional E-polarized field structures. CSI–EPT, on the other hand, does not exploit any particular field structure and can be extended to the vectorial three-dimensional case, turning CSI–EPT into a volumetric EPT reconstruction method.

We carried out several comparisons between reconstructions obtained with 2D CSI–EPT, foIC-EPT, and 3D CSI–EPT. Our simulations show that care needs to be exercised when a 2D reconstruction approach is followed or reconstruction artefacts are obtained in the reconstructed dielectric tissue profiles. Specifically, we showed that, by using 2D methods, erroneous reconstructions could be obtained, since the longitudinal variations of the transverse vector potential are completely ignored in the data model for the $\hat{B}_1^+$ field. Moreover, vector potential itself is computed differently in 2D and 3D since longitudinal invariance is assumed in the 2D case. In fact, the transverse electric field and the longitudinal magnetic field vanish in 2D as a consequence of the (assumed) invariance of the object and external sources along the longitudinal direction. In 3D, however, all components of the electromagnetic field are present and their contributions to the measured data and object equations have to be taken into account. Of course, in some situations, an E-polarized field structure may be present in the midplane of a birdcage coil, but the scattered $\hat{B}_1^+$-field is also influenced by longitudinal variations of transverse vector potential $\partial_z \hat{A}_z$. These equations can only be simplified to 2D if we can guarantee that longitudinal invariance or the smoothness of certain field components can be imposed before any reconstruction algorithm is applied to the measured data. Therefore, cylindrical body parts, such as legs or arms, could be reliably reconstructed via 2D CSI–EPT, but this at least requires further validation through simulations and measurements using cylindrical phantom models with known dielectric characteristics.

No assumptions on the fields are imposed in 3D CSI–EPT and reconstruction errors due to such assumptions are therefore avoided. Moreover, 3D CSI–EPT is a volumetric reconstruction method, and is not restricted to a specific plane within the configuration. Reliable reconstructions can be obtained within any desired domain of interest provided that $\hat{B}_1^+$ data are available within this domain. Unfortunately, computation times significantly increase when applying 3D CSI–EPT. Depending on the number of unknowns in the EPT reconstruction problem, 3D CSI–EPT may take many iterations to converge to the desired error tolerance, with total computation times of hours or days, even on dedicated computers or servers. In future research, we focus on accelerating the convergence rate of 3D CSI–EPT by including preconditioning techniques in CSI–EPT (as described in Reference [23], for example) that exploit all a priori knowledge we have about the object or body part that needs to be reconstructed. This knowledge can also be used to construct an accurate initial guess, thereby possibly further accelerating CSI–EPT.

In our experiments, we used simulated $\hat{B}_1^+$-field data to test the performance of 2D and 3D CSI–EPT on strongly inhomogeneous structures, and to study the differences between two- and three-dimensional CSI–EPT approaches. In real-world measurements, the data obviously differ from simulated data, and CSI–EPT should be adapted so that it can handle measured $\hat{B}_1^+$ data. In this respect, we identified three practical issues that need to be addressed, which are part of our current CSI–EPT research.

First, in practice, the $\hat{B}_1^+$-field is obtained in polar form through separate amplitude and phase measurements. In both cases, the collected data are contaminated with noise. Therefore, filtering or regularization techniques that suppress the effects of noise should be incorporated in CSI–EPT. Initial studies show that filtering of the data allows us to handle measured data in foIC-EPT [20] and, as demonstrated in Reference [14], total-variation (TV) regularization may suppress noise effects in CSI–EPT. However, due to the many possible choices for the regularization parameter in this method, it is presently not clear for which parameter or parameter range the TV CSI–EPT scheme is most effective.

Second, the phase that is measured in practice is not the phase of the $\hat{B}_1^+$-field, but the so-called transceive-phase from which the $\hat{B}_1^+$-phase can be extracted. To this end, the tranceive-phase approximation is often applied, but the validity of this approximation is not fully understood, and may lead to reconstruction errors in conductivity and permittivity profiles [28]. Fortunately, it is shown in Reference [31] that improved $\hat{B}_1^+$ phase approximations can be obtained from the tranceive phase by incorporating an iterative-phase correction scheme in the CSI–EPT reconstruction algorithm. This correction scheme seems to reliably retrieve $\hat{B}_1^+$ phase maps from the measured tranceive phase, and leads to improved conductivity and permittivity reconstructions compared with reconstructions that are obtained when tranceive-phase approximation is applied. We will include this phase-correction mechanism in future CSI–EPT implementations as well. Another option is to opt for phaseless approaches as, for example, proposed in References [32,33].

Finally, in practice, the current densities in the transmit coil that generate the incident field depend on the present object, and we must account for this loading effect as well. Specifically, integral representations for the fields in CSI–EPT are obtained using scattered-field formalism in which it is assumed that current density in the transmitting antenna is impressed and independent of the scatterer that may be present in the configuration. In practice, however, these currents do depend on the object, and, consequently, care must be taken when we compute the background field in CSI–EPT. One approach is, therefore, to simulate this loading effect using a suitable coil and body model in a commercial field solver and to extract current densities in the coil from this solver. The background field in CSI–EPT (the field without any load) can then be computed using these extracted currents. In this way, the loading effect encountered in practice can be incorporated in our CSI–EPT reconstruction algorithm.

Our final aim is, of course, to turn CSI–EPT into a practical reconstruction method to obtain accurate and reliable conductivity and permittivity tissue maps of an interior part of the human body at MR operation frequencies. Reconstruction results based on simulated data are very promising, and we think that, by addressing the practical issues discussed above, we will indeed make significant progress towards a reliable EPT reconstruction method that provides us with accurate dielectric tissue maps in practice.

Author Contributions: Conceptualization, R.R.; methodology, R.R.; software, R.L., P.F., W.B. and R.R.; validation, R.L and P.F.; formal analysis, R.R.; investigation, R.L., P.F., W.B. and R.R; resources, R.L., P.F., W.B., A.W. and R.R.; data curation, R.L., P.F., W.B. and R.R.; writing—original draft preparation, R.R., R.L. and P.F.; writing—review and editing, R.L., P.F., W.B., A.W. and R.R.; visualization, R.L. and P.F.; supervision, R.R. and A.W.; project administration, R.R. and A.W.; funding acquisition, R.R. and A.W.

Funding: The research of R. Leijsen was funded by the European Research Council Advanced NOMA MRI under grant number 670629. The research of P. Fuchs was funded in part through a collaboration between the Delft University of Technology and the Indian Institute of Science.

Acknowledgments: The authors thank Nico van den Berg of the Imaging Division of the University Medical Center Utrecht for many valuable discussions and support.

References

1. Hartwig, V. Engineering for safety assurance in MRI: Analytical, numerical and experimental dosimetry. *Magn. Reson. Imaging* **2015**, *33*, 681–689. [CrossRef] [PubMed]
2. Lagendijk, J.J.W. Hyperthermia treatment planning. *Phys. Med. Biol.* **2000**, *45*, R61–R76. [CrossRef] [PubMed]
3. Katscher, U.; van den Berg, C.A.T. Electric properties tomography: Biochemical, physical and technical background, evaluation and clinical applications. *NMR Biomed.* **2017**, *30*, 1–15. [CrossRef] [PubMed]
4. Gabriel, S.; Lau, R.W.; Gabriel, C. The dielectric properties of biological tissues: II. Measurements in the frequency range 10 Hz to 20 GHz. *Phys. Med. Biol.* **1996**, *41*, 2251–2269. [CrossRef] [PubMed]
5. Brown, B.H. Electrical impedance tomography (EIT): A review. *J. Med. Eng. Technol.* **2003**, *27*, 97–108. [CrossRef] [PubMed]
6. Chandra, R.; Zhou, H.; Balasingham, I.; Narayanan, R. On the Opportunities and Challenges in Microwave Medical Sensing and Imaging. *IEEE Trans. Biomed. Eng.* **2015**, *62*, 1667–1682. [CrossRef]
7. Zhang, X.; Liu, J.; He, B. Magnetic-Resonance-Based Electrical Properties Tomography: A Review. *IEEE Rev. Biomed. Eng.* **2014**, *7*, 87–96. [CrossRef]
8. Bal, L. Hybrid inverse problems and internal functionals. In *Inverse Problems and Applications: Inside Out II*; Uhlmann, G., Ed.; Cambridge University Press: New York, NY, USA, 2012; pp. 325–368, ISBN 978-1-107-03201-9.
9. Haacke, E.M.; Petropoulos, L.S.; Nilges, E.W.; Wu, D.H. Extraction of conductivity and permittivity using magnetic resonance imaging. *Phys. Med. Biol.* **1991**, *36*, 723–734. [CrossRef]
10. Katscher, U.; Voigt, T.; Findeklee, C.; Vernickel, P.; Nehrke, K.; Doessel, O. Determination of Electric Conductivity and Local SAR Via B1 Mapping. *IEEE Trans. Med. Imaging* **2009**, *28*, 1365–1374. [CrossRef]
11. Voigt, T.; Katscher, U.; Doessel, O. Quantitative Conductivity and Permittivity Imaging of the Human Brain Using Electric Properties Tomography. *Magn. Reson. Med.* **2011**, *66*, 456–466. [CrossRef]
12. Van Lier, A.L.H.M.W.; Brunner, D.O.; Pruessmann, K.P.; Klomp, D.W.J.; Luijten, P.R.; Lagendijk, J.J.W.; van den Berg, C.A.T. B1+ Phase Mapping at 7T and its Application for In Vivo Electrical Conductivity Mapping. *Magn. Reson. Med.* **2012**, *67*, 552–561. [CrossRef] [PubMed]
13. Hafalir, F.S.; Oran, O.F.; Gurler, N.; Ider, Y.Z. Convection-Reaction Equation Based Magnetic Resonance Electrical Properties Tomography (cr-MREPT). *IEEE Trans. Med. Imaging* **2014**, *33*, 777–793. [CrossRef] [PubMed]

14. Balidemaj, E.; van den Berg, C.A.T.; Trinks, J.; van Lier, A.L.H.M.W.; Nederveen, A.J.; Stalpers, L.J.A.; Crezee, H.; Remis, R.F. CSI-EPT: A Contrast Source Inversion Approach for Improved MRI-Based Electrical Properties Tomography. *IEEE Trans. Med. Imaging* **2015**, *34*, 1788–1796. 2015.2404944. [CrossRef] [PubMed]
15. Serrallés, J.E.C.; Daniel, L.; White, J.K.; Sodickson, D.K.; Lattanzi, R.; Polimeridis, A.G. Global Maxwell Tomography: A novel technique for electrical properties mapping based on MR measurements and volume integral equation formulations. In Proceedings of the 2016 IEEE International Symposium on Antennas and Propagation (APSURSI), Fajardo, Puerto Rico, 26 June–1 July 2016; pp. 1395–1396. [CrossRef]
16. Hong, R.; Li, S.; Zhang, J.; Zhang, Y.; Liu, N.; Yu, Z.; Liu, Q.H. 3-D MRI-Based Electrical Properties Tomography Using the Volume Integral Equation Method. *IEEE Trans. Microw. Technol.* **2017**, *65*, 4802–4811. [CrossRef]
17. Arduino, A.; Zilberti, L.; Chiampi, M.; Bottauscio, O. CSI-EPT in Presence of RF-Shield for MR-Coils. *IEEE Trans. Med. Imaging* **2017**, *36*, 1396–1404. [CrossRef] [PubMed]
18. Rahimov, A.; Litman, A.; Ferrand, G. MRI-based electric properties tomography with a quasi-Newton approach. *Inverse Probl.* **2017**, *33*, 105004. [CrossRef]
19. Gurler, N.; Ider, Y.Z. Gradient-Based Electrical Conductivity Imaging Using MR Phase. *Magn. Reson. Med.* **2017**, *77*, 137–150. [CrossRef]
20. Fuchs, P.S.; Mandija, S.; Stijnman, P.R.S.; Brink, W.M.; van den Berg, C.A.T.; Remis, R.F. First-Order Induced Current Density Imaging and Electrical Properties Tomography in MRI. *IEEE Trans. Comput. Imaging* **2018**, *4*, 624–631. [CrossRef]
21. Van den Berg, P.M.; Abubakar, A. Contrast source inversion method: State of art. *Prog. Electromagn. Res.* **2001**, *34*, 189–218. [CrossRef]
22. Van den Berg, P.M.; Kleinman, R.E. A contrast source inversion method. *Inverse Probl.* **1997**, *13*, 1607–1620. [CrossRef]
23. Van den Berg, P.M.; van Broekhoven, A.L.; Abubakar, A. Extended contrast source inversion. *Inverse Probl.* **1999**, *15*, 1325–1344. [CrossRef]
24. Van den Berg, P.M.; Haak, K.F.I. Profile inversion by error reduction in the source type integral equations. In *Wavefields and Reciprocity–Proceedings of a Symposium Held in Honour of Professor dr. A.T. de Hoop*; van den Berg, P.M., Blok, H., Fokkema, J.T., Eds.; Delft University Press: Delft, The Netherlands, 1996; pp. 87–98, ISBN 90-407-1402-9.
25. Van den Bergen, B.; Stolk, C.C.; van den Berg, J.B.; Lagendijk, J.J.W.; Van den Berg, C.A.T. Ultra fast electromagnetic field computations for RF multi-transmit techniques in high field MRI. *Phys. Med. Biol.* **2009**, *54*, 1253–1264. [CrossRef] [PubMed]
26. Leijsen, R.L.; Brink, W.M.; van den Berg, C.A.T.; Webb, A.G.; Remis, R.F. 3-D Contrast Source Inversion-Electrical Properties Tomography. *IEEE Trans. Med. Imaging* **2018**, *9*, 2080–2089. [CrossRef] [PubMed]
27. Christ, A.; Kainz, W.; Hahn, E.G.; Honegger, K.; Zefferer, M.; Neufeld, E.; Rascher, W.; Janka, R.; Bautz, W.; Chen, J.; et al. The Virtual Family–development of surface-based anatomical models of two adults and two children for dosimetric simulations. *Phys. Med. Biol.* **2010**, *55*, N23–N38. [CrossRef] [PubMed]
28. Van Lier, A.L.H.M.W.; Raaijmakers, A.; Voigt, T.; Lagendijk, J.J.W.; Luijten, P.R.; Katscher, U.; van den Berg, C.A.T. Electrical Properties Tomography in the Human Brain at 1.5, 3, and 7T: A Comparison Study. *Magn. Reson. Med.* **2014**, *71*, 354–363. [CrossRef] [PubMed]
29. Saad, Y. *Iterative Methods for Sparse Linear Systems*; SIAM: Philadelphia, PA, USA, 2003; ISBN 0-89871-534-2.
30. Balidemaj, E.; van den Berg, C.A.T.; van Lier, A.L.H.M.W.; Nederveen, A.J.; Stalpers, L.J.A.; Crezee, H.; Remis, R.F. B1-based SAR reconstruction using contrast source inversion–electric properties tomography (CSI-EPT). *Med. Biol. Eng. Comput.* **2017**, *55*, 225–233. [CrossRef] [PubMed]
31. Stijnman, P.R.S.; Mandija, S.; Fuchs, P.S.; Remis, R.F.; van den Berg, C.A.T. Transceive Phase Corrected Contrast Source Inversion-Electrical Properties Tomography. In Proceedings of the ISMRM Joint Annual Meeting, Paris, France, 15–18 June 2018; Volume 5087.

32. Arduino, A.; Bottauscio, O.; Chiampi, M.; Zilberti, L. Magnetic resonance-based imaging of human electric properties with phaseless contrast source inversion. *Inverse Probl.* **2018**, *34*, 084002. [CrossRef]
33. Bevacqua, M.T.; Bellizzi, G.G.; Isernia, T.; Crocco, L. A method for quantitative imaging of electrical properties of human tissues from only amplitude electromagnetic data. *Inverse Probl.* **2019**, *35*, 025006. [CrossRef]

Article

Novel Stopping Criteria for Optimization-Based Microwave Breast Imaging Algorithms

Cameron Kaye *, Ian Jeffrey and Joe LoVetri

Electrical and Computer Engineering, University of Manitoba, Winnipeg, MB R3T 5V6, Canada;
ian.jeffrey@umanitoba.ca (I.J.); Joe.LoVetri@umanitoba.ca (J.L.)
* Correspondence: umkayecj@myumanitoba.ca

Received: 25 April 2019; Accepted: 14 May 2019; Published: 22 May 2019

Abstract: A discontinuous Galerkin formulation of the Contrast Source Inversion algorithm (DGM-CSI) for microwave breast imaging employing a frequency-cycling reconstruction technique has been modified here to include a set of automated stopping criteria that determine a suitable time to shift imaging frequencies and to globally terminate the reconstruction. Recent studies have explored the use of tissue-dependent geometrical mapping of the well-reconstructed real part to its imaginary part as initial guesses during consecutive frequency hops. This practice was shown to improve resulting 2D images of the dielectric properties of synthetic breast models, but a fixed number of iterations was used to halt DGM-CSI inversions arbitrarily. Herein, a new set of stopping conditions is introduced based on an intelligent statistical analysis of a window of past iterations of data error using the two-sample Kolmogorov-Smirnov (K-S) test. This non-parametric goodness-of-fit test establishes a pattern in the data error distribution, indicating an appropriate time to shift frequencies, or terminate the algorithm. The proposed stopping criteria are shown to improve the efficiency of DGM-CSI while yielding images of equivalent quality to assigning an often liberally overestimated number of iterations per reconstruction.

Keywords: breast imaging; microwave imaging; discontinuous Galerkin method (DGM); contrast source inversion (CSI); stopping criteria; Kolmogorov-Smirnov (K-S) test

1. Introduction

Microwave imaging (MWI) has made steady progress towards widespread clinical application in breast cancer detection and monitoring over the past decade, offering a safety advantage over established x-ray-based modalities due to its use of non-ionizing radiation and a significant cost benefit over magnetic resonance imaging (MRI). While its currently attainable spatial resolution does not match that of cancer screening tools like mammography, concurrent use of MWI of the breast in its current state could lead to decreased false positive rates among certain population groups, particularly pregnant women or those with radiographically dense ("Category C" and "Category D") breasts, as classified by the American College of Radiology's Breast Imaging Reporting and Data System (BI-RADS) [1,2].

Methods of utilizing MWI for biomedical applications have been investigated for almost half a century [3,4]. Some recent representative research can be found in [5]. MWI algorithms can be split into those producing qualitative images, where the reconstructed value at each pixel is a relative quantity, and those producing quantitative reconstructions of a physical property, typically the complex-valued permittivity. Examples of qualitative algorithms are those utilized in so-called radar imaging techniques [6], and those retrieving only qualitative aspects of the target, such as its support [7,8]. The interest in this paper are quantitative imaging methods that solve the full electromagnetic inverse problem.

It is well known that the electromagnetic inverse problem associated with MWI, where one attempts to reconstruct the complex-valued permittivity of an object-of-interest (OI), is ill-posed [9]. The ill-posedness stems from the non-uniqueness of the inverse-source problem where one attempts to reconstruct the electromagnetic sources responsible for a remotely measured field. In the electromagnetic inverse problem, these sources are the contrast-sources that are illuminated by the interrogating field. Various regularization techniques have been developed over several decades of research to deal with the ill-posedness of the problem [10]. The electromagnetic inverse problem is not only ill-posed but also non-linear with repect to the two unknowns within the inaccessible imaging domain, the electromagnetic field and the permittivity. The Contrast Source Inversion (CSI) technique effectively linearizes the problem by casting the data-error norm in terms of contrast-source variables, which vary with the interrogating field, and regularization is performed by introducing the so-called Maxwellian regularizer which is written in terms of the contrast-sources and the contrast [11]. Much research has been performed on advancing the CSI algorithm since it was first reported, including the addition of a multiplicative regularizer [12].

For biomedical imaging applications, which are of interest in the context of the present work, the incorporation of discretized numerical inversion models into CSI that are based on either the finite-element method (FEM) or the discontinuous Galerkin method (DGM) forward solvers have provided much flexiblity and several advantages for the overall inversion process [13–15]. Most importantly, these forward solvers have allowed one to incorporate prior information into the inversion model that has the effect of regularizing the inversion, enabling high-contrast reconstructions such as are required for breast imaging [16–19]. The flexibility of these partial differential equation (PDE) based solvers has also allowed advancements in data-acquisition systems utilized to acquire scattered-field data for MWI [20,21]. Other potential advancements in MWI for breast tumor detection include the use of contrast agents [22]

Recent work in multi-frequency imaging has shown promise in increasing the spatial resolution and correspondingly, the amount of distinguishable anatomic detail in MWI reconstructions. For instance, while the use of "frequency hopping" has long been shown to be an effective means of obtaining images from high-frequency data by feeding low-frequency solutions as "initial guesses" into inversion algorithms [23], a so-called "frequency cycling" and "tissue-dependent mapping" framework has been adopted in recent studies to perform multi-frequency imaging demonstrating significant enhancement in the quality of resulting 2D images [24,25]. In those works, an arbitrary fixed number of iterations was used as the stopping condition for the imaging algorithm, but it was noted that further exploration into more intelligent stopping criteria was justified, particularly based on the observation that running the algorithm longer than necessary could deteriorate the imaging results. In fact, to date, there has been little exploration into optimizing the conditions to which a microwave imaging algorithm should adhere when determining the appropriate point to move on from intermediate inversions to the next frequency in a multi-frequency sequence, or when to terminate reconstructions entirely.

While monitoring the convergence of data error has been the most common objective approach to halting microwave imaging algorithms, as it is often difficult to ascertain whether a chosen error limit will undershoot, overshoot, or consistently meet the attainable image reconstruction quality for a given dataset, several early studies focusing on synthetic and experimental imaging performance fell into the same habit of simply pre-assigning an arbitrary number of iterations for the algorithm to complete, chosen for convenience or through trial and error [11,26–28].

Although there have been previous attempts to develop more intelligent stopping criteria for imaging algorithms in recent literature [17], a more thorough statistical analysis of the data error is undertaken here and employed as part of a logical multi-variable framework of stopping criteria for a frequency-cycling tissue-dependent mapping formulation of Contrast Source Inversion (CSI). For comparison purposes, some 2D imaging results from a recent study are included in this work to demonstrate the significant gains in efficiency and image quality granted through the incorporation of the described stopping criteria [25]. Reconstruction results are subjected to quantitative analysis using common image error metrics to provide an objective measurement of these improvements.

2. Materials and Methods

2.1. DGM-CSI Algorithm

An implementation of CSI using a high-order frequency-domain formulation of Maxwell's curl equations has been used for all 2D breast images presented here, employing the discontinuous Galerkin method (DGM) as a forward solver. Nodal coefficients in high-order polynomial expansions represent unknown field and contrast quantities in the resulting DGM-CSI algorithm. Independent from this choice of implementation, among the benefits of CSI is its use of operators that are functions of frequency and the material properties of the background medium only, which do not change from iteration to iteration [11]. As aforementioned, further improvement in performance has been demonstrated through the use of a tissue-dependent mapping process and the practice of cycling back to low-frequency reconstructions [25], and a brief overview of this process is provided in Section 2.2. Further details specific to DGM-CSI can be found in [14,29], but a summary of the relevant quantities and definitions involved in the CSI cost functional is included here, since it pertains to the discussion of data and domain error that follows.

The microwave imaging problem in Figure 1 depicts an object of interest (OI) with an unknown dielectric contrast $\chi(\vec{r})$, a function of position vector $\vec{r}$, typically defined as

$$\chi(\vec{r}) = \frac{\hat{\varepsilon}_r(\vec{r}) - \hat{\varepsilon}_b}{\hat{\varepsilon}_b}, \tag{1}$$

where $\hat{\varepsilon}_r$ is the complex relative permittivity and $\hat{\varepsilon}_b$ is the complex background permittivity. In breast imaging applications, the background permittivity is usually represented as the homogeneous immersion medium in which the breast is submerged and the imaging domain $\mathcal{D}$ is an area typically chosen to be contained by the outer skin layer of the breast. An array of $t = 1, 2, \ldots, T$ transmitters is used to generate probing incident fields; the resulting scattered fields are measured by an array of R receivers that make up a discrete measurement surface $\mathcal{S}$.

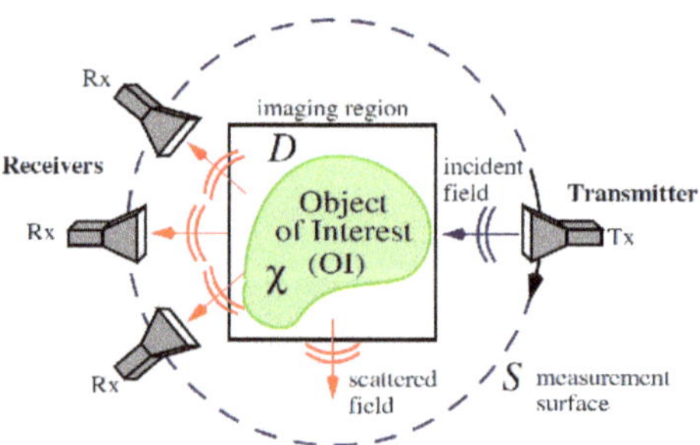

Figure 1. Two-dimensional representation of the imaging problem. The object of interest (OI) has an unknown contrast χ, where $\mathcal{D}$ is the imaging domain and $\mathcal{S}$ is the surface containing the transmitters (Tx) and receivers (Rx).

It is assumed that the reader is familiar with the standard CSI algorithm, and only the aspects important to the proposed stopping criteria will be reviewed. The CSI algorithm solves for the contrast $\chi(\vec{r})$ of breast tissue within $\mathcal{D}$ from this scattered field data by minimizing the following two-part cost functional:

$$\mathcal{F}^{\text{CSI}}(\chi, w_t) = \mathcal{F}^{\mathcal{S}}(w_t) + \mathcal{F}^{\mathcal{D}}(\chi, w_t), \tag{2}$$

where both the contrast $\chi(\vec{r})$ and the transmitter-dependent contrast sources $w_t(\vec{r}) = \chi(\vec{r})u_t^{\text{tot}}$ (a product of the contrast and the total field u_t^{tot}) are unknown and confined to the domain $\mathcal{D}$.

The first term of Equation (2) represents the "data error", given by

$$\mathcal{F}^{\mathcal{S}}(w_t) = \frac{\sum_t \left\| u_t^{\text{sct}} - \mathcal{L}_{\mathcal{S}}\{w_t\} \right\|_{\mathcal{S}}^2}{\sum_t \left\| u_t^{\text{sct}} \right\|_{\mathcal{S}}^2}, \tag{3}$$

and correspondingly, the second term represents the "domain error",

$$\mathcal{F}^{\mathcal{D}}(\chi, w_t) = \frac{\sum_t \left\| \chi u_t^{\text{inc}} - w_t + \chi \mathcal{L}_{\mathcal{D}}\{w_t\} \right\|_{\mathcal{D}}^2}{\sum_t \left\| \chi u_t^{\text{inc}} \right\|_{\mathcal{D}}^2}, \tag{4}$$

with $\|\cdot\|_{\mathcal{D}}$ and $\|\cdot\|_{\mathcal{S}}$ denoting the L_2 norms on $\mathcal{D}$ and $\mathcal{S}$. The measured scattered field data is represented above by u_t^{sct} at R receiver locations per transmitter t, and the incident field inside $\mathcal{D}$ is u_t^{inc}. The forward operator $\mathcal{L}_{\mathcal{S}}$ converts contrast source estimates w_t in $\mathcal{D}$ to scattered field values at R receiver points per transmitter t on the measurement surface $\mathcal{S}$; $\mathcal{L}_{\mathcal{D}}$ performs a similar function, but transforms contrast sources to scattered field values within $\mathcal{D}$. For further simplicity, the "data error" per iteration i can be defined from the numerator of Equation (3) as

$$\rho_t^{(i)} = d_t - \mathcal{L}_{\mathcal{S}}\{w_t^{(i)}\} \tag{5}$$

and correspondingly, the "domain error" from the numerator of Equation (4) as

$$r_t^{(i)} = \chi^{(i)} u_t^{\text{inc}} - w_t^{(i)} + \chi^{(i)} \mathcal{L}_{\mathcal{D}}\{w_t^{(i)}\}. \tag{6}$$

These error values, particularly the data error of Equation (5), are subject to analysis for evaluating convergence behaviour of DGM-CSI in this study, described in Section 2.3.

2.2. Frequency-Cycling Tissue-Dependent Mapping Technique

The well-documented dielectric properties of breast tissues lead to the observation that the geometry of the real and imaginary parts of tissues' permittivity profiles should be qualitatively similar [24,25]. The quantitative permittivity values of different types of breast tissue also form reasonably discrete ranges that are highly correlated between real and imaginary components, making it easy to stratify these values into expected tissue types (i.e., fat, transitional, fibroglandular, and cancer) [30]. Through a tissue-dependent mapping of the imaginary part based on geometric and quantitative reconstructions of the dielectric constant, an "artificial" imaginary component was proposed as part of the initial guess for the next frequency dataset in the imaging sequence, to good effect [24,25].

In addition to tissue-dependent mapping, this previous work invoked "frequency cycling" as opposed to conventional frequency hopping, where frequency cycling simply continues the reconstruction process past the highest frequency in a multi-frequency sequence by returning to the lowest frequency dataset (with an initial guess based on the highest frequency solution) [25]. The initial studies using DGM-CSI have shown that this practice preserves the fine detail acquired from high-frequency data, and produces better results than a single incremental frequency-hopping sequence operating for the same total number of iterations. In a practical microwave imaging scenario without the benefit of foreknowledge of this total number of iterations, however, a potential drawback to frequency cycling would be the increased number of iterations required to produce images when arbitrary or overly-stringent stopping conditions are applied.

2.3. Stopping Criteria for Single-Frequency Reconstructions

As described in Section 2.1, iterative inversion methods such as DGM-CSI attempt to solve the inverse scattering problem by minimizing the cost functional of Equation (2), which includes a data error $\rho_t^{(i)}$ (Equation (5)) represented by the difference between the measured and computed fields on $\mathcal{S}$ for a given transmitter t. If U is the total number of data points across all receivers R and sources T, then the $U \times 1$ data-error vector $\bar{\rho}^{(i)}$ can be represented as a concatenation of every transmitter's data errors, such that

$$\bar{\rho}^{(i)} = [\rho_1^{(i)}; \rho_2^{(i)}; \ldots; \rho_T^{(i)}] \tag{7}$$

After all contrast sources $w_t^{(i)}$ are updated for each transmitter t, the inversion process by convention performs a convergence check that calculates the new data-error vector $\bar{\rho}^{(i+1)}$. The goal is to reduce this new data error as much as possible, and aside from simply assigning a maximum number of iterations, one could monitor the data error until it falls below an arbitrarily-chosen tolerance level δ such that some form of a termination condition examining its sum, maximum value, or norm applies, commonly implemented as

$$\text{if}\,(\|\bar{\rho}^{(i+1)}\| < \delta)\,\text{stop}.$$

Alternatively, the difference between the normalized data-error vector at two successive iterations, $\bar{\rho}^{(i)}$ and $\bar{\rho}^{(i+1)}$, can be calculated until it falls within a prescribed tolerance to check for convergence [26].

Such conditions would seem to be reasonable criteria by which termination of the inversion algorithm should take place. However, the total data error often has little bearing on the accuracy of the resulting reconstruction, despite properly converging to a preset minimum. While liberally large values of error tolerance will logically terminate the imaging algorithm prematurely, it has also been observed that conservatively small choices of error tolerance will result in a poorer quality image. This outcome is likely due to the supposition that the minimization of a cost functional towards zero arguably means accounting for any and all error in modeling, calibration, and measurements (i.e., noise) through non-physical changes in the reconstructions. The apparent inadequacy of having the data error converge to within an arbitrary tolerance alone served as motivation to develop a multifaceted approach to stopping criteria that improves the reconstructions of a frequency-cycling reconstruction technique through statistical analysis of data and domain error of each iteration.

This proposed framework stops the imaging algorithm based primarily on the statistical *distribution* of the contributing terms to the data-error vector $\bar{\rho}$. Referring the reader back to Equation (7), assuming a best-case scenario where the forward operator $\mathcal{L}_\mathcal{S}$ is able to perfectly model the field behaviour of the imaging system and the contrast sources w_t had been iteratively solved to the exact solution, the data error ρ_t would be reduced to the difference between the field values d_t experimentally measured at the receivers for all given transmitters and the ideal noise-free field values generated by $\mathcal{L}_\mathcal{S}\{w_t^{(i)}\}$. This difference would amount to the noise of the experimental system, which would presumably converge to a known pattern, such as a Gaussian distribution.

However, it is clear that even if one could acquire completely noise-free measurement data, it is not practical, or even possible, to find a numerical model that could exactly match it. This so-called modeling error, attributed to mismatches between the approximate representation of the imaging environment $\mathcal{L}_\mathcal{S}$ (which varies by formulation, mesh granularity, order number, etc.) and the true, unknowable functions that perfectly predict field behaviours within the system, invalidates the assumption that the remaining data-error vector $\bar{\rho}$ (upon convergence of the imaging algorithm to a solution) consists simply of noise. Modeling error is the primary reason why data calibration must be performed on experimental data, and it is not a trivial issue to address in MWI [31].

Correspondingly, since the contributions to the data error are multifactorial in nature and vary with each experimental set-up and possibly with each target, it will not necessarily converge to a known family of statistical distributions. Use of any of the myriad of statistical normality tests available (such as the Anderson–Darling or Shapiro–Wilk test) on the data error as part of the proposed algorithmic stopping criteria would therefore be unreliable. However, the profile of the data error distribution (regardless of its final form) should reach a steady state when the algorithm has sufficiently converged to a particular solution. To confirm that this trend occurs during a reconstruction requires comparison of the current iteration's data error distribution to those of multiple previous iterations, which is accomplished in this work with an implementation of the *two-sample Kolmogorov-Smirnov* (K-S) test.

The two-sample K-S test is a non-parametric goodness-of-fit hypothesis test that evaluates the maximum absolute difference D between the cumulative distribution functions (CDF) of two sample data vectors over a range of x in each data set. Recall that the CDF (F_X) of a continuous random variable X can be expressed as the integral of its probability density function (PDF) f_x (that is, a function whose value at any given sample can be interpreted as providing a relative likelihood that the value of the random variable would be equal to that sample), such that

$$F_X(x) = \int_{-\infty}^{x} f_X(t)\, dt. \tag{8}$$

In the discrete case, $F_X(x)$ would be equivalent to the proportion of values in $f_X(x)$ less than or equal to a specified value of x.

Suppose the data-error vector PDFs from two consecutive iterations of the DGM-CSI imaging algorithm, $\bar{\rho}^{(i)}$ and $\bar{\rho}^{(i+1)}$, each have some observed CDF, say $\hat{P}^{(i)}(x)$ and $\hat{P}^{(i+1)}(x)$; that is, for any specified value of x, the value of $\hat{P}(x)$ is the proportion of error values less than or equal to x in the corresponding dataset. Consider the quantity

$$D = \max_{x}(|\hat{P}^{(i+1)}(x) - \hat{P}^{(i)}(x)|), \tag{9}$$

that represents the maximum error between two CDFs. The K-S test makes use of this D value to return a probability, in the form of a scalar asymptotic p-value in the range of $[0, 1]$, which represents the likelihood that the two samples were drawn from the same distribution. More precisely, the p-value is defined as the probability of observing a test statistic (D) as extreme or more extreme than the observed value, for the null hypothesis that the data in these two vectors are from the same continuous distribution [32–35]. In the implementation of the test, the decision to reject the null hypothesis is based on comparing the returned p-value with a preset significance level (commonly 5% for many applications). The test becomes very accurate for large sample sizes, and is believed to be reasonably accurate for sample sizes U_1 and U_2, such that $(U_1 \times U_2)/(U_1 + U_2) \geq 4$, which is the case for a collection of data error values in a system of 24 transmitters and receivers operating at a single frequency (representing a total of 576 data points).

While the strength of the two-sample K-S test lies in determining when two samples are from differing distributions, the nature of the test guarantees that the smaller the value of D, the higher the returned p-value will be. This feature allows the test's p-value to be used as a numerical indicator for how likely the two samples were drawn from the same distribution (i.e., how closely their CDFs match), as opposed to its primary use in *rejecting* the null hypothesis.

Empirical studies of the real and imaginary parts of the data error of DGM-CSI reconstructions indeed revealed that the shape of the probability distribution curves fluctuate significantly with large variances in early iterations (Figure 2) but gradually reach a steady state in later iterations as expected, once all remarkable subjective change in the resulting images had ceased (Figure 3). Since any two neighboring iterations' data error may only change slightly and be interpreted as being drawn from the same distribution by the two-sample K-S test, it was decided that a sliding window of past iterations' data errors should be examined and compared to that of the current iteration. Moreover, the threshold value that the returned p-value should surpass for any given K-S test within this sliding window needed to be high enough to similarly reject subtler variations in two iterations of data error that would otherwise pass the goodness-of-fit test. This wider breadth of analysis ensured that more gradual changes were detected so that any corresponding stopping criteria would not halt the reconstruction prematurely.

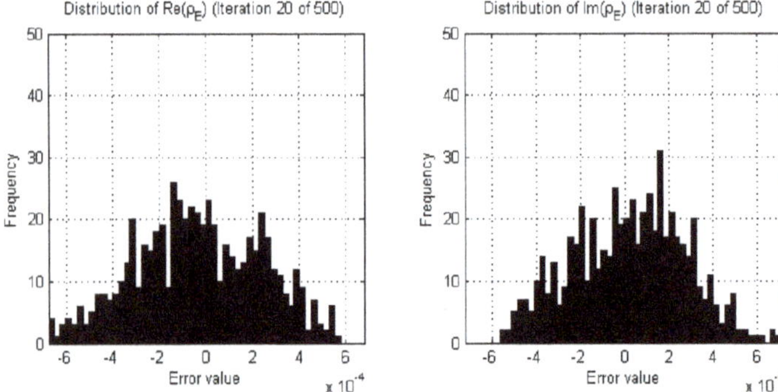

Figure 2. Examples of histograms demonstrating estimates of the probability distribution of the real and imaginary parts of the data error (ρ_E) during a DGM-CSI inversion of an arbitrary 2D synthetic breast model early in the reconstruction process (iteration 20).

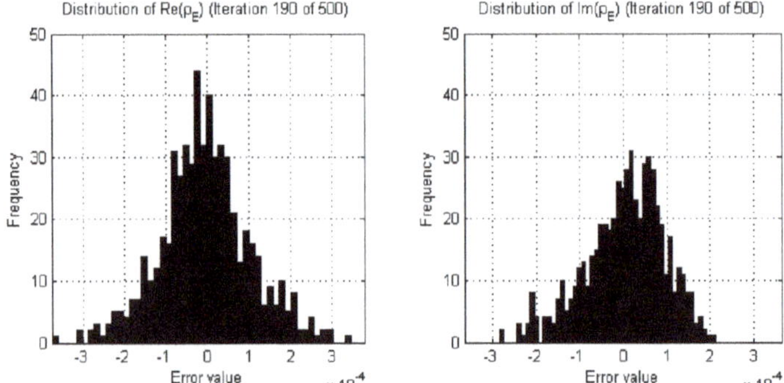

Figure 3. Examples of histograms demonstrating estimates of the probability distribution of the real and imaginary parts of the data error (ρ_E) during the same DGM-CSI inversion as Figure 2, later in the reconstruction process (iteration 190).

Therefore, the window size of iterations under scrutiny, the p-value threshold (assumed constant over the window), and the percentage of that window's iterations reaching or surpassing that p-value threshold, are all parameters to the stopping criteria for the algorithm based on the K-S test, which is applied separately for the real and imaginary parts of each iteration's data error. A preliminary study attempting to establish the effects of these parameters was undertaken for the frequency-cycled reconstruction of a 2D breast model, described in Section 3.1.

It should be noted that the statistical analysis governing the aforementioned stopping criteria only makes use of the *data* error, which may be sufficient for certain MWI algorithms (such as the distorted Born iterative method) whose minimized cost functional consists solely of a data error term. However, as shown in Section 2.1, CSI-based iterative inversion algorithms make use of cost functionals containing a second domain error term dealing with total field values inside the imaging domain [11,26]. Therefore, another ad hoc stopping condition based on the domain error is also prudently employed in the algorithm, as described in the next subsection.

2.4. Global Termination of Multi-Frequency Reconstructions

The stopping conditions described in Section 2.3 are appropriate for halting single-frequency inversions, but multiple-frequency reconstructions have thus far not been addressed. In a frequency-hopping scenario, the stopping criteria based on K-S tests alone can be easily employed to determine when each individual inversion should terminate and move on to the next frequency, with global termination occurring following reconstruction of the highest frequency.

However, in a frequency-cycling reconstruction scheme, the cycle of reconstructions could theoretically continue *ad infinitum* without another explicit set of rules in place governing the global termination of the imaging process. A separate, secondary ad hoc global termination criterion was therefore implemented, based on a previous study [17]; once each frequency's dataset in the cycle is subjected to the K-S test stopping criteria at least once, if the percentage of relative change in domain error between two successive iterations falls below 0.1%, then it is deemed appropriate to globally terminate the frequency cycle and end the reconstruction. This provision could also serve as a fallback contingency in the event that the data error-based conditions turn out to be too strict to satisfy for a given individual frequency.

2.5. Full Description of Multi-Frequency Imaging Procedure

The key concepts involved in the proposed multi-frequency imaging procedure, while implemented herein for the DGM-CSI algorithm, can easily be applied (with minor modifications) to other iterative algorithms widely used in MWI, including Gauss-Newton Inversion (GNI) and distorted Born Iterative Method (DBIM). As such, a descriptive approach to the technique is undertaken here in order to keep its applicability as general as possible. Taking into account the stopping criteria described throughout Section 2.3 and using the procedural description of the frequency-cycling reconstruction approach with tissue-dependent mapping described in [25], the new consolidated multi-frequency imaging procedure is proposed as follows:

1. The real and imaginary parts of the complex permittivity are reconstructed using the lowest frequency data available (e.g., 1.0 GHz). The termination point of this reconstruction is dictated by the results of successive two-sample K-S tests performed on both the real and imaginary parts of the data error separately, comparing the current iteration to those of a sliding window of past iterations, governed by a choice of parameters for p-value, window size, and percentage of windowed iterations reaching this p-value threshold. For robustness, a back-up termination condition may be implemented, either related to the relative change in domain error for CSI-based algorithms as described in Section 2.4, or a maximum number of iterations.

2. A point-by-point search through the reconstructed real part of each nodal basis coefficient in the DGM-CSI mesh (or more generally, each mesh element or pixel of the reconstructed image) classifies the type of breast tissue. This classification is based solely on the range of expected values of dielectric constant at that frequency, as outlined in [25].

3. An initial guess for the next imaging frequency (e.g., 2.0 GHz) is generated using the tissue-dependent mapping process [24,25]. It consists of the unmodified real parts of the reconstructed ε_r at the mesh nodal points, and a new imaginary part created from a simple linear interpolation of the expected range of dielectric loss values, based on the appropriate Cole-Cole models of tissues classified in Step 2. This technique preserves the geometry of the real and imaginary parts of the solution.

4. This new initial guess for the complex permittivity is used to run the inversion algorithm at the next frequency (e.g., 2.0 GHz). As per the procedure outlined in [25], the user may choose to keep the imaginary part constant during this inversion and update only the real part to converge to a new solution. This "anchoring" process has been shown to improve overall imaging results due to the tendency of CSI-based inversion algorithms to cause significant deterioration of the imaginary

part at high-frequency reconstructions. Again, the aforementioned parameterized stopping criteria would be primarily employed to determine the appropriate point to halt this reconstruction.

5. If more than two frequencies are used in the frequency hop, steps 2–4 are repeated as necessary until the reconstruction of the final frequency of the succession is complete (e.g., 3.0 GHz). This succession may include "frequency cycling"; that is, returning the inversion algorithm to the lowest frequency data and incrementally stepping through each frequency again [25].

If a simple frequency-hopping scheme is employed, the reconstruction will terminate after the highest-frequency inversion, again as decided by the parameterized stopping criteria based on K-S tests of the data error. However, if a frequency-cycling scheme is used, and if the imaginary part of the solution has been "anchored" in place following the first inversion in Step 1, there are further considerations that come into play for global termination of the imaging process, and to address concerns of full CSI optimization:

- When each available dataset in the frequency cycle has been used *at least once* to contribute to the overall image reconstruction, a global termination criterion will become active, which will monitor the relative change in the domain error between successive iterations (Section 2.4). If this relative change falls below 0.1% at any point, the current reconstruction is halted and the frequency cycle is broken.
- Regardless of the frequency at which the algorithm was halted by this relative domain error threshold, if the imaginary part of the solution has been continuously held constant during the frequency cycle after Step 1, one last initial guess is generated as in Step 3 and a final reconstruction is run at the lowest frequency available (e.g., 1.0 GHz) with both the real and imaginary parts allowed to converge to a solution (i.e., the imaginary part is no longer "anchored"). This final inversion is terminated by the parameterized stopping criteria *or* a relative change of domain error between successive iterations falling below 0.1%, whichever occurs first. The purpose of this final run is to demonstrate the stability of the final solution and ensure that its imaginary part, despite being originally based on the geometry and tissue properties of the real part, does indeed satisfy full CSI optimization.

2.6. Synthetic Breast Models

Initial testing of the described stopping criteria and comparison to fixed-iteration scenarios were carried out on synthetic transverse magnetic (TM) data collected from a two-dimenstional MRI-derived BI-RADS Category C (heterogeneously dense) 2D breast model supplied by the University of Calgary, derived from an MRI slice of a cancer patient with a breast tumour visible at the "3 o'clock" position. To demonstrate the robustness of the technique across different breast phantoms, an additional two-dimensional MRI-derived BI-RADS Category D ("extremely dense") healthy breast model supplied by the University of Wisconsin's public database (ID: 070604PA2) was employed. Both models are depicted at three different frequencies (1.0 GHz, 2.0 GHz, and 3.0 GHz) in Figures 4 and 5, illustrating the similarities between breast tissues' dielectric permittivity at the low-gigahertz band. The values of the tissue-dependent complex permittivity in all cases were calculated for every frequency using an appropriate fitted single-pole Cole-Cole model (or equivalently, a single-pole Debye model as the exponent parameter $\alpha = 0$), and for the University of Calgary model, then subjected to random perturbations of $\pm 10\%$ [36].

Data was collected for both models in a low-loss background of $\epsilon_r = 23 - 1.13i$ with 24 transmitters and 24 receivers evenly distributed at a radius of 10 cm, employing a finite-element method (FEM) forward solver with a finely-discretized mesh independent from that used for the DGM-CSI inversion. The synthetic data was collected for open boundary conditions at 1.0 GHz, 2.0 GHz, and 3.0 GHz. To test the robustness of the stopping criteria to problems with PEC boundaries, and to better reflect the capabilities of a recent University of Manitoba imaging system prototype [37], data was collected a second time with the Category C model surrounded by a circular PEC boundary with a radius of 15 cm at a reduced bandwidth. The frequencies used for the PEC-bounded case were

1.0 GHz, 1.25 GHz, and 1.5 GHz. Uniformly-distributed noise at 5% of the maximum field magnitude was used to corrupt the E_z scattered electric field data for the Category C model, and several noise levels (3%, 5%, 7.5%, and 10%) were tested for the Category D model to evaluate the effect of noise on the stopping criteria's performance. The skin thickness and inner skin boundary both remained unknown; the only prior information used during inversions was the outer skin boundary.

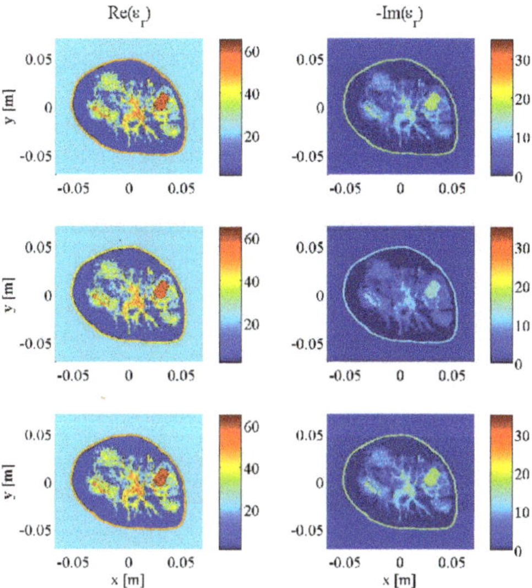

Figure 4. Complex dielectric properties of University of Calgary 2D synthetic breast model at 1.0 GHz (**top**), 2.0 GHz (**middle**) and 3.0 GHz (**bottom**).

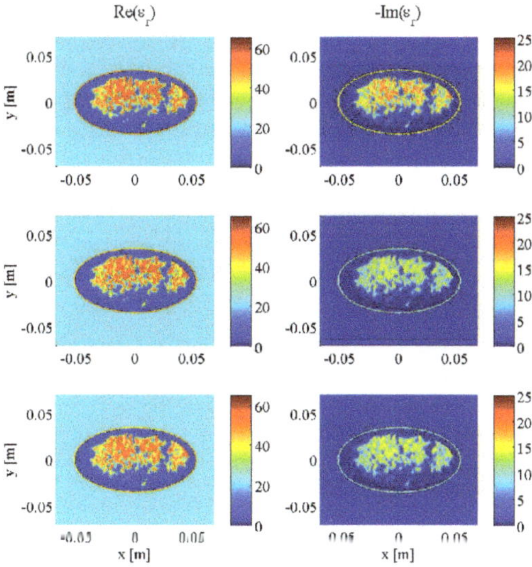

Figure 5. Complex dielectric properties of University of Wisconsin 2D synthetic breast model at 1.0 GHz (**top**), 2.0 GHz (**middle**) and 3.0 GHz (**bottom**).

2.7. Error Calculation

An objective method of evaluating the DGM-CSI algorithm's performance for differing imaging scenarios was accomplished by comparing the reconstruction results to the original Category C and Category D synthetic models of the *current* frequency being used in the reconstruction cycle, unless otherwise specified. The models and reconstructions each employed different underlying meshes, so their complex permittivity values were interpolated onto a common 250×250-pixel square grid ($N = 62{,}500$) that encompassed an area covering $(x_1, y_1) = (-0.07, -0.07)$ [m] to $(x_2, y_2) = (0.07, 0.07)$ [m]. The relative error of L_1 and L_2 norms of the difference between the model (actual) complex permittivity $\hat{\varepsilon}_r^{\,\text{act}}$ and the reconstructed complex permittivity $\hat{\varepsilon}_r^{\,\text{rec}}$ over each pixel k were used as primitives to evaluate the performance of the imaging algorithm. For convenience, these will hereafter be referred to as the relative error norms REN_1 and REN_2, and are calculated and expressed as percentages, where

$$
\begin{aligned}
\text{REN}_1(\hat{\varepsilon}_r^{\,\text{act}}, \hat{\varepsilon}_r^{\,\text{rec}}) \;&=\; \frac{\|\hat{\varepsilon}_r^{\,\text{act}} - \hat{\varepsilon}_r^{\,\text{rec}}\|_1}{\|\hat{\varepsilon}_r^{\,\text{act}}\|_1} \times 100\% \\[2mm]
&=\; \frac{\sum\limits_{k=1}^{N} |\hat{\varepsilon}_r^{\,\text{act}}(k) - \hat{\varepsilon}_r^{\,\text{rec}}(k)|}{\sum\limits_{k=1}^{N} |\hat{\varepsilon}_r^{\,\text{act}}(k)|} \times 100\%,
\end{aligned}
\tag{10}
$$

$$
\begin{aligned}
\text{REN}_2(\hat{\varepsilon}_r^{\,\text{act}}, \hat{\varepsilon}_r^{\,\text{rec}}) \;&=\; \frac{\|\hat{\varepsilon}_r^{\,\text{act}} - \hat{\varepsilon}_r^{\,\text{rec}}\|_2}{\|\hat{\varepsilon}_r^{\,\text{act}}\|_2} \times 100\% \\[2mm]
&=\; \sqrt{\frac{\sum\limits_{k=1}^{N} [\hat{\varepsilon}_r^{\,\text{act}}(k) - \hat{\varepsilon}_r^{\,\text{rec}}(k)]^2}{\sum\limits_{k=1}^{N} [\hat{\varepsilon}_r^{\,\text{act}}(k)]^2}} \times 100\%.
\end{aligned}
\tag{11}
$$

The raw numerical value of the norms will be slightly reduced since the square grid covers an area *outside* of the imaging domain of the problem, which will consist of common unaltered background permittivity in both the model and reconstruction and thus represent pixels of zero error. Since the REN values serve primarily as a means of monitoring iterative trends in solution convergence and to demonstrate comparative improvement between imaging scenarios with equally-sized interpolated grids and imaging domains, the impact of this artificial error reduction is of little significance.

3. Results and Discussion

3.1. Imaging with Open Boundaries

As mentioned in the introduction of the stopping criteria in Section 2.3, the approach had three variables: the p-value threshold that any given two-sample K-S test would use to conclude the two data error samples were drawn from the same distribution, the window size of past iterations' data errors compared to that of the current iteration, and the percentage of that window's iterations that would need to to reach or surpass the chosen p-value threshold in order to halt the reconstruction at the current frequency, deemed the "pass percentage". For the same Category C breast model in Figure 4, DGM-CSI was run in a frequency-cycled configuration using the rules described in Section 2.5 for every combination of p-value thresholds of 0.90, 0.95 and 0.99, window sizes of 10, 30, and 50, and pass percentages of 80%, 90%, and 100%. There were 27 simulations whose total number of iterations would vary based on the laxity or stringency of stopping criteria associated with low and high variable values respectively. Error norms (REN_1 and REN_2) for each scenario were plotted for every frequency jump

(Figure 6), and the best variable combination (resulting in the lowest error metric magnitudes upon termination) was found to be a *p*-value threshold of 0.99, a window size of 30, and a pass percentage of 80%. This combination of values was subsequently used for all future simulations employing the stopping criteria. Note that in Figure 6 and in all subsequent plots of REN values, the *x*-axis' number of iterations represents the cumulative total throughout all frequency-hopping or frequency-cycling steps of the reconstruction.

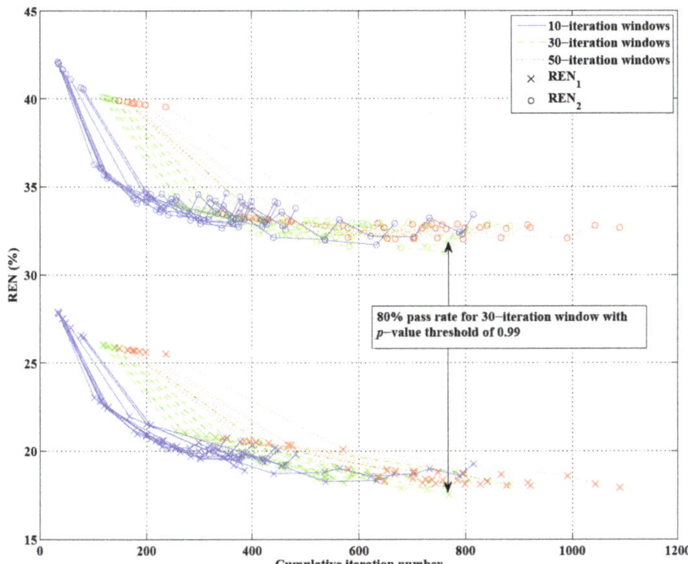

Figure 6. Relative error norms of open-boundary 2D DGM-CSI reconstructions across several choices of stopping criteria parameter values. Each curve in the figure corresponds to a frequency-cycled inversion for a particular choice of the three parameters (Section 3.1), with the window sizes coded by color for convenience. Data points on each curve correspond to REN values for the reconstructed model at each frequency change. Arrows point to the final REN values of simulations that terminated with the lowest relative error norms, indicating the best combination of parameters among those tested for this imaging scenario.

Having established an appropriate set of variables to govern the stopping criteria, a series of simulations adhering to the full description of the technique in Section 2.5 could be carried out to demonstrate its benefit in efficiency over conventional methods. The REN values of three scenarios (A, B, and C from Table 1) comparing fixed-iteration reconstructions (with and without tissue-dependent mapping) to the use of both tissue-dependent mapping and the stopping criteria, are shown in Figure 7. The final imaging reconstructions of these three cases are also depicted in Figure 8. As the proposed stopping conditions introduce an unpredictable halt to the reconstruction cycle that could occur at any frequency and also includes a mandatory return to the lowest frequency (in this case, 1.0 GHz) for the final optimization, each REN data point may not represent consecutive frequencies of the cycle. For consistency, the REN values shown in Figure 7 are therefore all calculated based on the same 2.0 GHz model from Figure 4. For clarity, the frequency associated with each data point plotted in Figure 7, along with the triggering condition that caused a change in frequency or termination in the reconstruction (when relevant), is shown in Table 1.

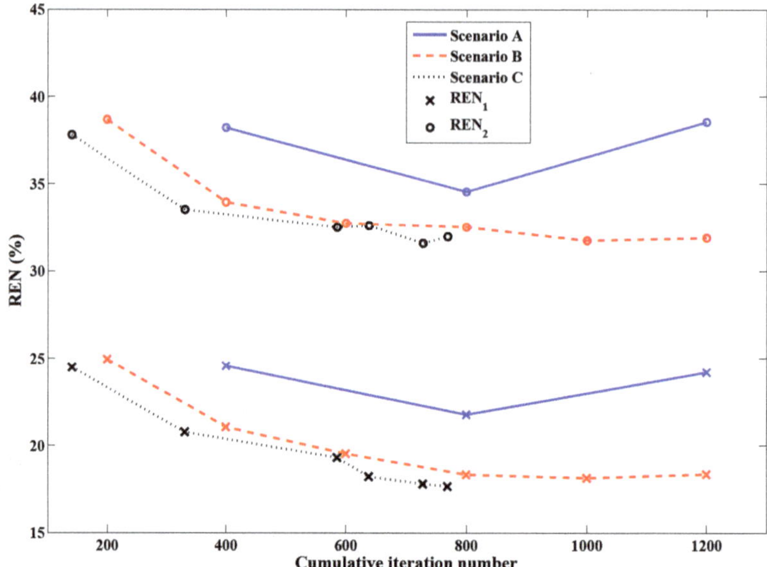

Figure 7. Relative error norms of frequency-hopping and frequency-cycled reconstructions: Scenario A (solid line)—without tissue-dependent mapping at 400 iterations per frequency terminating after first inversion of 3.0 GHz, Scenario B (dashed line)—with tissue-dependent mapping at 200 iterations per frequency, cycling through reconstruction frequencies once (with imaginary component "anchored" following initial 1.0 GHz inversion), Scenario C (dotted line)—with tissue-dependent mapping and stopping criteria in place, terminating after two consecutive inversions of 1.0 GHz data (one with the imaginary component "anchored" and the final run with the imaginary component freely optimized according to the guidelines of Section 2.5). See Table 1 for further details.

Table 1. Reconstruction progression of open boundary scenarios (Figure 7).

Scenario	TM	SC	Frequency:						Total
			No. of Iterations (Stopping Condition)						
			Components Reconstructed						
			1.0 GHz:	2.0 GHz:	3.0 GHz:				
A	No	No	400 (F)	400 (F)	400 (F)				1200
			Re, Im	*Re, Im*	*Re, Im*				
			1.0 GHz:	2.0 GHz:	3.0 GHz:	1.0 GHz:	2.0 GHz:	3.0 GHz:	
B	Yes	No	200 (F)	200 (F)	200 (F)	200 (F)	200 (F)	200 (F)	1200
			Re, Im	*Re*	*Re*	*Re*	*Re*	*Re*	
			1.0 GHz:	2.0 GHz:	3.0 GHz:	1.0 GHz:	2.0 GHz:	1.0 GHz:	
C	Yes	Yes *	140 (KS)	190 (KS)	255 (KS)	53 (KS)	89 (DE)	41 (KS)	768
			Re, Im	*Re*	*Re*	*Re*	*Re*	*Re, Im*	

TM = tissue mapping; SC = stopping criteria; F = fixed; KS = Kolmogorov-Smirnov test window; DE = Domain error; *Re* = Real part; *Im* = Imaginary part. * 80% of a 30-iteration data error window must reach or surpass a K-S test *p*-value threshold of 0.99.

Although DGM-CSI, like other iterative microwave imaging algorithms, performs reasonably well in recovering the real component of breast models it reconstructs, it has difficulty producing an accurate profile of the imaginary part, most notably at higher frequencies. This deficiency is illustrated from the Scenario A inversion of the Category C breast model in the top image of Figure 8. Subjectively, the benefit of employing the tissue-dependent mapping in the fidelity of the imaginary component is obvious in both the middle and bottom images. It has been already shown elsewhere that the described mapping technique and "anchoring" of the imaginary component at each subsequent frequency stage, along with the practice of frequency cycling, imparts a benefit to the quality of the real part of the reconstructions following the same number of iterations [24,25]. However, these other studies only tested reconstructions that used fixed numbers of iterations as termination conditions; the addition of the stopping criteria for the latter case here (Scenario C) does not appreciably degrade the quality of the final image, while reducing the total number of iterations by over 35%.

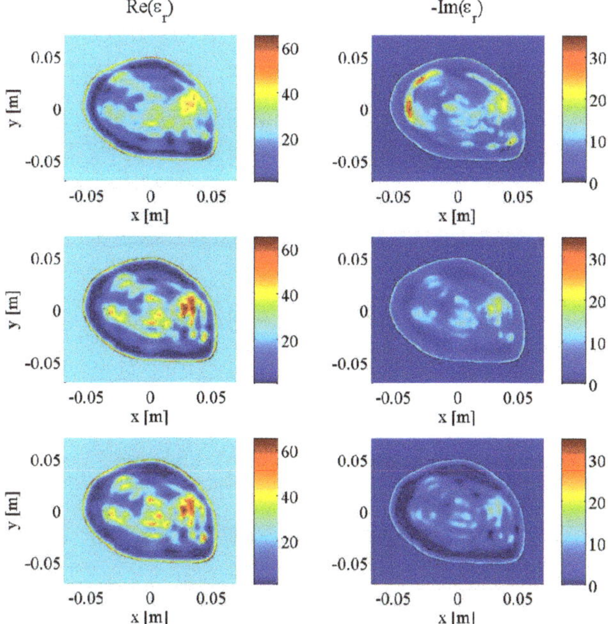

Figure 8. Final results of DGM-CSI frequency-hopping and frequency-cycled complex dielectric property reconstruction of synthetic breast model using 1.0–3.0 GHz data, without modification of intermediate initial guesses (Scenario A—**top**), using tissue-dependent mapping at fixed 200 iterations per frequency (Scenario B—**middle**), and employing stopping criteria and tissue-dependent mapping (Scenario C—**bottom**).

3.2. Imaging with PEC Boundaries

A set of simulations similar to those carried out in the previous section were performed using PEC boundaries and data from frequencies of 1.0 GHz, 1.25 GHz and 1.50 GHz, to demonstrate this method's robustness to smaller frequency bandwidths and different boundary conditions, as mentioned in Section 2.6. Scenarios D and E are imaging simulations recreated from an earlier study that employed fixed-iteration reconstructions exclusively [25]; they are included in order to emphasize the benefit of adding stopping criteria in Scenario F. The imaging results are shown in Figure 9 with a breakdown of the reconstruction progression documented in Table 2. A plot of the relative error norms again similar to the open boundary cases, employing REN calculations based on the original 1.25 GHz model, is also provided in Figure 10.

As explained in [25], it was expected that the results would not be as impressive as open-boundary scenarios since PEC-bounded problems are generally more difficult to solve, and data was taken at lower frequencies. However, it is again clear from Figure 9 that there is an improvement in the recovery of the imaginary component when tissue-dependent mapping is used (middle and bottom images), more so for Scenario F. A modest improvement in the real part of ε_r for Scenarios E and F is also noted. Without the use of stopping criteria, however, Scenario E overshoots the expected values of dielectric constant in several areas. Its REN values depicted in Figure 10 correspondingly suffer increases relating to the fact that the reconstruction cycle has perhaps been allowed to run too long, as pointed out in [25]. The introduction of the stopping criteria has effectively prevented this degradation by terminating the reconstruction cycle close to the observed nadir of these error values from Scenario E, while once again demonstrating a sizable increase in computational efficiency by reducing the number of iterations by an impressive 66%.

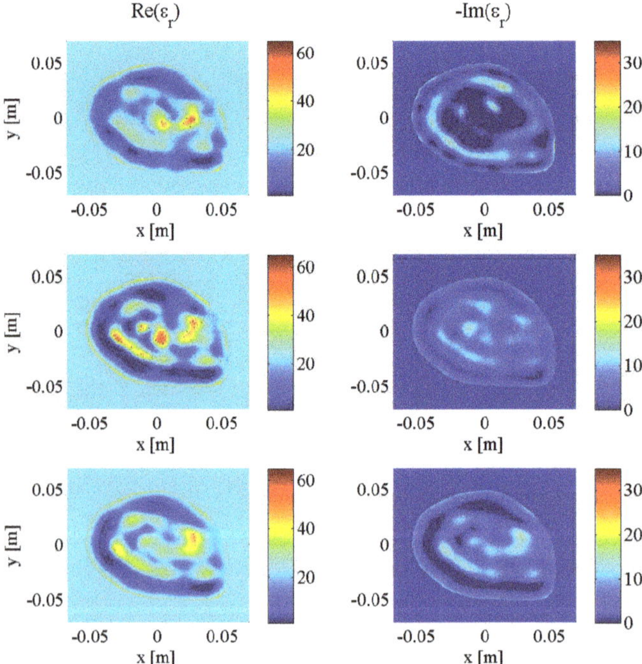

Figure 9. Final results of DGM-CSI frequency-hopping and frequency-cycled complex dielectric property reconstruction of PEC-bounded synthetic breast model using 1.0–1.5 GHz data, without modification of intermediate initial guesses (Scenario D—**top**), using tissue-dependent mapping at fixed 200 iterations per frequency (Scenario E—**middle**), and employing stopping criteria and tissue-dependent mapping (Scenario F—**bottom**).

Table 2. Reconstruction progression of PEC-bounded scenarios (Figure 10).

Scenario	TM	SC	Frequency:						Total
			No. of Iterations (Stopping Condition)						
			Components Reconstructed						
			1.0 GHz:	**1.25 GHz:**	**1.5 GHz:**				
D	No	No	400 (F)	400 (F)	400 (F)				1200
			Re, Im	*Re, Im*	*Re, Im*				
			1.0 GHz:	**1.25 GHz:**	**1.5 GHz:**	**1.0 GHz:**	**1.25 GHz:**	**1.5 GHz:**	
E	Yes	No	200 (F)	200 (F)	200 (F)	200 (F)	200 (F)	200 (F)	1200
			Re, Im	*Re*	*Re*	*Re*	*Re*	*Re*	
			1.0 GHz:	**1.25 GHz:**	**1.5 GHz:**	**1.0 GHz:**	**1.0 GHz:**		
F	Yes	Yes *	123 (KS)	124 (KS)	97 (KS)	14 (DE)	48 (DE)		406
			Re, Im	*Re*	*Re*	*Re*	*Re, Im*		

TM = tissue mapping; SC = stopping criteria; F = fixed; KS = Kolmogorov-Smirnov test window; DE = Domain error; *Re* = Real part; *Im* = Imaginary part. * 80% of a 30-iteration data error window must reach or surpass a K-S test *p*-value threshold of 0.99.

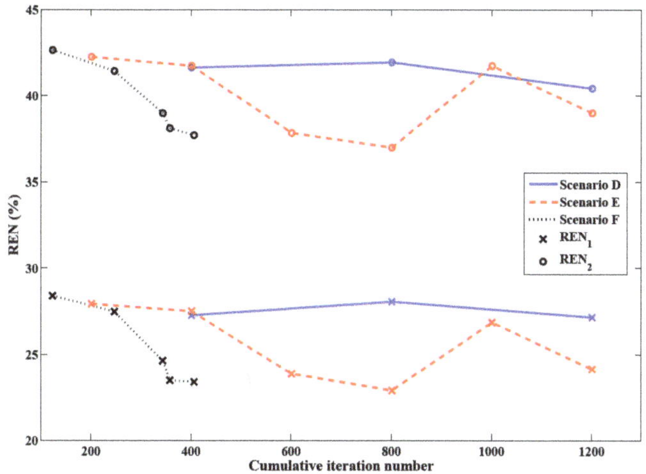

Figure 10. Relative error norms of reconstructions with PEC boundaries: Scenario D (solid line)—without tissue-dependent mapping at 400 iterations per frequency terminating after first inversion of 1.5 GHz data, Scenario E (dashed line)—with tissue-dependent mapping at 200 iterations per frequency, cycling through reconstruction frequencies once (with imaginary component "anchored" following initial 1.0 GHz inversion), Scenario F (dotted line)—with tissue-dependent mapping and stopping criteria in place, terminating after two consecutive inversions of 1.0 GHz data (one with the imaginary component "anchored" and the final run with the imaginary component freely optimized according to the guidelines of Section 2.5). See Table 2 for further details.

3.3. Effect of Noise Levels

To validate that the stopping criteria developed and tested on the BI-RADS Category C 2D model would perform similarly on different breast models and be robust to varying levels of noise, another full set of simulations were carried out on the aforementioned Category D model from the University of Wisconsin database using open boundary data at 1.0 GHz, 2.0 GHz, and 3.0 GHz (Figure 5). The uniformly-distributed noise used to corrupt the E_z scattered electric field data was generated at

3%, 5%, 7.5%, and 10% of the maximum field magnitude. The imaging results are shown in Figure 11 with a breakdown of the reconstruction progression documented in Table 3. A plot of the relative error norms again similar to the Category C cases, employing REN calculations based on the original 2.0 GHz Category D model, is also provided in Figure 12.

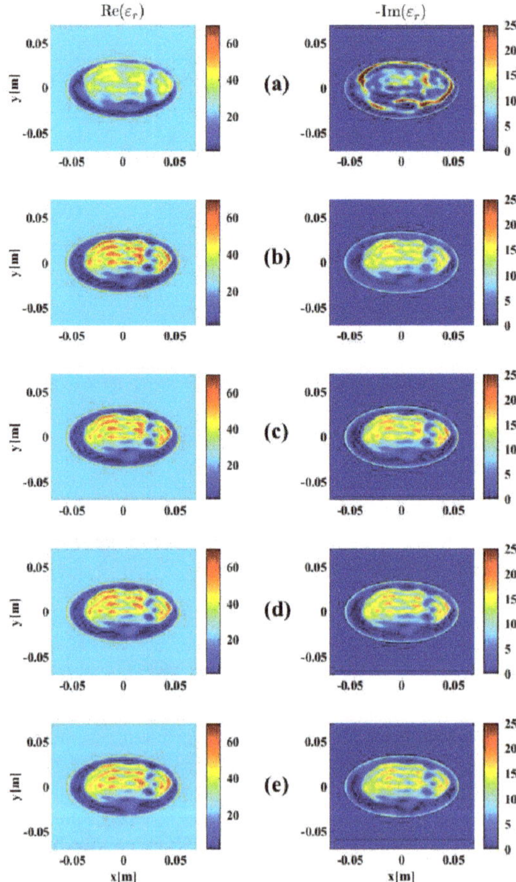

Figure 11. Final results of DGM-CSI frequency-hopping and frequency-cycled complex dielectric property reconstruction of open-boundary Category D synthetic breast model using 1.0–3.0 GHz data: (**a**) Scenario G—without tissue-dependent mapping at 400 iterations per frequency terminating after first inversion of 3.0 GHz (5% noise), (**b**) Scenario H—with tissue-dependent mapping and stopping criteria in place at 3% noise, (**c**) Scenario I—with tissue-dependent mapping and stopping criteria in place at 5% noise, (**d**) Scenario J—with tissue-dependent mapping and stopping criteria in place at 7.5% noise, (**e**) Scenario K—with tissue-dependent mapping and stopping criteria in place at 10% noise. See Table 3 for further details.

As shown in the final images of Figure 11, there is an obvious improvement using tissue-dependent mapping and stopping criteria over the simple frequency-hopping approach (Scenario G), especially in the imaginary part, whose geometry is not reconstructed properly and contains a large overshooting boundary artifact. However, remarkably, there is very little subjective difference in the images reconstructed from data corrupted with different levels of noise (from 3% to 10% for Scenarios H through K), with subtle changes in the real and imaginary parts visible only with careful scrutiny.

The relative error norms of all scenarios depicted in Figure 12 and the reconstruction break-down in Table 3 offer more objective measurements of the effect of noise. Specifically, lower final REN values are observed for images inverted from data containing lower noise levels, as would be expected. Also, the stopping criteria transitions frequencies and halts the entire reconstruction cycle sooner for higher noise levels. These trends are consistent for all reconstructions employing tissue-dependent mapping and stopping criteria in each level of noise used in the simulations. For instance, data containing 3% noise (Scenario H) runs longest, for a total of 759 iterations; this total is observed to gradually decrease at the 5% and 7.5% noise levels, and data containing 10% noise (Scenario K) runs for only 448 iterations.

These observations are congruous with the theory introduced in Section 2.3 that data containing higher magnitudes of noise would have less useful information for the optimization algorithm to extract before the distribution of data error reached a steady state (i.e., before the algorithm began to reconstruct noise) and continuing the inversion would be superfluous, or perhaps even detrimental to the final image. Overall, while these additional 2D simulations do not necessarily represent a fully comprehensive investigation across all possible breast models and imaging scenarios, the performance of the proposed stopping criteria appears satisfactory on the Category D model, and behaves as expected on noisier datasets, even with parameter choices derived from the Category C model reconstructions.

Table 3. Reconstruction progression of Category D model noise-variant scenarios (Figure 12).

Scenario (Noise %)	TM	SC	Frequency:						Total
			No. of Iterations (Stopping Condition)						
			Components Reconstructed						
			1.0 GHz:	2.0 GHz:	3.0 GHz:				
G (5%)	No	No	400 (F)	400 (F)	400 (F)				1200
			Re, Im	*Re, Im*	*Re, Im*				
			1.0 GHz:	2.0 GHz:	3.0 GHz:	1.0 GHz:	2.0 GHz:	1.0 GHz:	
H (3%)	Yes	Yes *	155 (KS)	140 (KS)	265 (KS)	50 (KS)	52 (DE)	43 (DE)	759
			Re, Im	*Re*	*Re*	*Re*	*Re*	*Re, Im*	
			1.0 GHz:	2.0 GHz:	3.0 GHz:	1.0 GHz:	1.0 GHz:		
I (5%)	Yes	Yes *	139 (KS)	140 (KS)	265 (KS)	34 (DE)	8 (DE)		586
			Re, Im	*Re*	*Re*	*Re*	*Re, Im*		
			1.0 GHz:	2.0 GHz:	3.0 GHz:	1.0 GHz:	1.0 GHz:		
J (7.5%)	Yes	Yes *	91 (KS)	112 (KS)	234 (KS)	37 (DE)	13 (DE)		487
			Re, Im	*Re*	*Re*	*Re*	*Re, Im*		
			1.0 GHz:	2.0 GHz:	3.0 GHz:	1.0 GHz:	1.0 GHz:		
K (10%)	Yes	Yes *	87 (KS)	90 (KS)	228 (KS)	11 (DE)	32 (DE)		448
			Re, Im	*Re*	*Re*	*Re*	*Re, Im*		

TM = tissue mapping; SC = stopping criteria; F = fixed; KS = Kolmogorov-Smirnov test window; DE = Domain error; *Re* = Real part; *Im* = Imaginary part. * 80% of a 30-iteration data error window must reach or surpass a K-S test *p*-value threshold of 0.99.

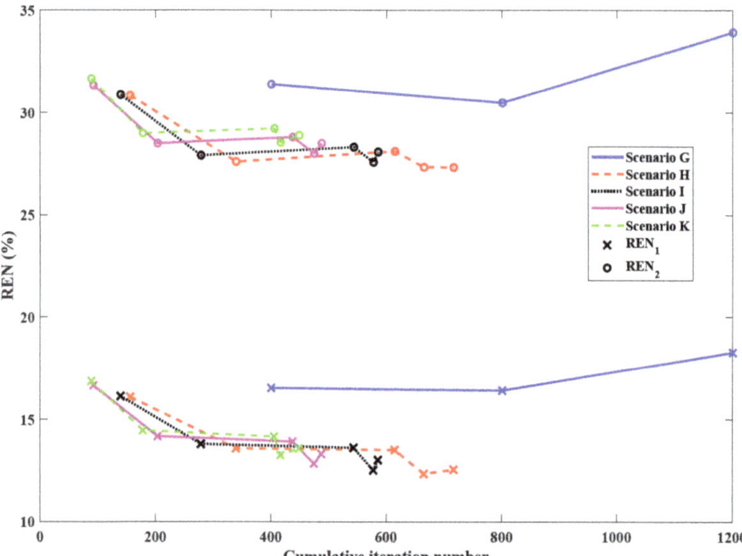

Figure 12. Relative error norms of frequency-hopping and frequency-cycled reconstructions of Category D model: Scenario G (blue solid line)—without tissue-dependent mapping at 400 iterations per frequency terminating after first inversion of 3.0 GHz (5% noise), Scenario H (red dashed line)—with tissue-dependent mapping and stopping criteria in place at 3% noise, Scenario I (black dotted line)—with tissue-dependent mapping and stopping criteria in place at 5% noise, Scenario J (magenta solid line)—with tissue-dependent mapping and stopping criteria in place at 7.5% noise, Scenario K (green dashed line)—with tissue-dependent mapping and stopping criteria in place at 10% noise. See Table 3 for further details.

4. Conclusions

It has been demonstrated that a stopping condition based on the similarities of the statistical distribution of data error at successive iterations has the capability of significantly speeding up the DGM-CSI algorithm while having no appreciable detrimental effect on the final imaging results, even yielding superior solutions than fixed-iteration scenarios. This boost of efficiency is granted through novel use of sequential two-sample K-S tests on the data error to determine appropriate times to halt the imaging algorithm, truncating likely unnecessary (and in some cases counterproductive) reconstruction iterations.

Whether or not the best set of parameters used in this analysis represents the true optimal values for the stopping criteria for other breast models, or if these same values are applicable to 3D inversions, is open to debate. However, given the promising imaging results obtained from this initial set of tests, and the gains in efficiency provided through significant reduction of the number of iterations needed to converge to these solutions, it was judged appropriate to keep the best parameters from this introductory exploration of the method and leave a more comprehensive examination of parameter optimization for future investigations.

The most interesting form of the proposed multi-frequency imaging procedure employs tissue-dependent mapping and "anchoring" of the imaginary part along with the new stopping criteria. This ad hoc procedure also returns the reconstruction cycle to stable low-frequency data after recovering anatomical detail from higher frequencies, allowing both real and imaginary parts to properly converge before global termination. The advantages gained by this use of both tissue-dependent mapping and novel stopping criteria have also been demonstrated to be robust for different breast models, frequency bandwidths, boundary conditions, and noise levels.

In fact, the stopping criteria have a more dramatic effect on the imaging results obtained from the more complex PEC-bounded problem and data with higher noise levels, suggesting this technique may be especially useful for avoiding image degradation and reconstruction artifacts in difficult cases such as full 3D reconstructions in semi-resonant enclosures, including experimental measurements that may be excessively noisy or contain poor frequency data unsuitable for inversion. To explore this possibility, preliminary extension of this technique to the 3D DGM-CSI algorithm developed at the University of Manitoba is a logical next step, along with parameter optimization studies of the stopping criteria for data from experimental systems.

Author Contributions: Conceptualization, C.K., I.J. and J.L.; Methodology, C.K., I.J. and J.L.; Resources, C.K., I.J. and J.L.; Software, C.K. and I.J.; Supervision, J.L.; Validation, C.K. and I.J.; Writing—original draft, C.K.; Writing—review & editing, C.K., I.J. and J.L.

Funding: This research is partially supported by the Natural Sciences and Engineering Research Council of Canada (NSERC).

Acknowledgments: The authors would like to thank the University of Calgary for providing the 2D Category C synthetic breast model used for the reconstructions presented in this work.

Conflicts of Interest: The authors declare no conflict of interest.

References

1. Trentham-Dietz, A.; Kerlikowske, K.; Stout, N.K.; Miglioretti, D.L.; Schechter, C.B.; Ergun, M.A.; van den Broek, J.J.; Alagoz, O.; Sprague, B.L.; van Ravesteyn, N.T.; et al. Tailoring breast cancer screening intervals by breast density and risk for women aged 50 years or older: Collaborative modeling of screening outcomes. *Ann. Intern. Med.* **2016**, *165*, 700–712. [CrossRef] [PubMed]
2. D'Orsi, C.J.; Mendelson, E.B.; Morris, E.A. (Eds.) *ACR BI-RADS Atlas: Breast Imaging Reporting and Data System*, 5th ed.; American College of Radiology: Reston, VA, USA, 2012.
3. Boerner, W.M.; Brand, H.; Cram, L.A.; Gjessing, D.T.; Jordan, A.K.; Keydel, W.; Schwierz, G.; Vogel, M. (Eds). *Inverse Methods in Electromagnetic Imaging*; D. Reidel Publishing: Dordrecht, The Netherlands, 1985.
4. Larsen, L.E.; Jacobi, J.H. (Eds.) *Medical Applications of Microwave Imaging*; IEEE Press: New York, NY, USA, 1986.
5. Kosmas, P.; Crocco, L. Introduction to Special Issue on "Electromagnetic Technologies for Medical Diagnostics: Fundamental Issues, Clinical Applications and Perspectives". *Diagnostics* **2019**, *9*, 19. [CrossRef]
6. Sill, J.M.; Fear, E.C. Tissue sensing adaptive radar for breast cancer detection—Experimental investigation of simple tumor models. *IEEE Trans. Microw. Theory Tech.* **2005**, *53*, 3312–3319. [CrossRef]
7. Delbary, F.; Brignone, M.; Bozza, G.; Aramini, R.; Piana, M. A visualization method for breast cancer detection using microwaves. *SIAM J. Appl. Math.* **2010**, *70*, 2509–2533. [CrossRef]
8. Cakoni, F; Colton, D.; Monk, P. Qualitative Methods in Inverse Electromagnetic Scattering Theory: Inverse Scattering for Anisotropic Media. *IEEE Antennas Propag. Mag.* **2017**, *59*, 24–33. [CrossRef]
9. Colton, D.; Kress, R. *Inverse Acoustic and Electromagnetic Scattering Theory*; Springer Science & Business Media: Berlin, Germany, 2012; Volume 93.
10. Engl, H.W.; Hanke, M.; Neubauer, A. *Regularization of Inverse Problems*; Kluwer Academic Publishers: Dordrecht, The Netherlands, 1996.
11. Van den Berg, P.M.; Kleinman, R.E. A contrast source inversion method. *Inverse Probl.* **1997**, *13*, 1607–1620. [CrossRef]
12. Abubakar, A.; Van den Berg, P.M.; Mallorqui, J.J. Imaging of biomedical data using a multiplicative regularized contrast source inversion method. *IEEE Trans. Microw . Theory Tech.* **2002**, *50*, 1761–1771. [CrossRef]
13. Zakaria, A.; Gilmore, C.; LoVetri, J. Finite-element contrast source inversion method for microwave imaging. *Inverse Probl.* **2010**, *26*, 115010. [CrossRef]
14. Jeffrey, I.; Geddert, N.; Brown, K.; LoVetri, J. The Time-Harmonic Discontinuous Galerkin Method as a Robust Forward Solver for Microwave Imaging Applications. *Prog. Electromagn. Res.* **2015**, *154*, 1–21. [CrossRef]

15. Brown, K.G.; Geddert, N.; Asefi, M.; LoVetri, J.; Jeffrey, I. Hybridizable Discontinuous Galerkin Method Contrast Source Inversion of 2-D and 3-D Dielectric and Magnetic Targets. *IEEE Trans. Microw. Theory Tech.* **2019**, *67*, 1766–1777. [CrossRef]

16. Baran, A.; Kurrant, D.; Zakaria, A.; Fear, E.; LoVetri, J. Breast Imaging Using Microwave Tomography with Radar-Based Tissue-Regions Estimation. *Prog. Electromagn. Res.* **2014**, *149*, 161–171. [CrossRef]

17. Kurrant, D.J.; Baran, A.; Fear, E.C.; LoVetri, J. Integrating prior information into microwave tomography Part 1: Impact of detail on image quality. *Med. Phys.* **2017**, *44*, 6461–6481. [CrossRef]

18. Kurrant, D.; Baran, A.; LoVetri, J.; Fear, E. Integrating prior information into microwave tomography part 2: Impact of errors in prior information on microwave tomography image quality. *Med. Phys.* **2017**, *44*, 6482–6503. [CrossRef]

19. Abdollahi, N.; Kurrant, D.; Mojabi, P.; Omer, M.; Fear, E.; LoVetri, J. Incorporation of Ultrasonic Prior Information for Improving Quantitative Microwave Imaging of Breast. *IEEE J. Multiscale Multiphys. Comput. Tech.* **2019**, *4*, 98–110. [CrossRef]

20. Nemez, K.; Baran, A.; Asefi, M.; LoVetri, J. Modeling Error and Calibration Techniques for a Faceted Metallic Chamber for Magnetic Field Microwave Imaging. *IEEE Trans. Microw. Theory Tech.* **2007**, *65*, 4347–4356. [CrossRef]

21. Asefi, M.; Baran, A.; LoVetri, J. An Experimental Phantom Study for Air-Based Quasi-Resonant Microwave Breast Imaging. *IEEE Trans. Microw . Theory Tech.* **2019**. [CrossRef]

22. Shea, J.D.; Kosmas, P.; Van Veen, B.D.; Hagness, S.C. Contrast-enhanced microwave imaging of breast tumors: A computational study using 3D realistic numerical phantoms. *Inverse Probl.* **2010**, *26*, 074009. [CrossRef]

23. Chew, W.; Lin, J. A frequency-hopping approach for microwave imaging of large inhomogeneous bodies. *IEEE Microw. Guided Wave Lett.* **1995**, *5*, 439–441. [CrossRef]

24. Kaye, C.; Jeffrey, I.; LoVetri, J. Enhancement of multi-frequency microwave breast images using a tissue-dependent mapping technique with discontinuous Galerkin contrast source inversion. In Proceedings of the AES 2016—4th Advanced Electromagnetics Symposium, Malaga, Spain, 26–28 July 2016; pp. 37–39.

25. Kaye, C.; Jeffrey, I.; LoVetri, J. Improvement of Multi-Frequency Microwave Breast Imaging through Frequency Cycling and Tissue-Dependent Mapping. *IEEE Trans. Antennas Propag.* **2019**, submitted.

26. Abubakar, A.; van den Berg, P.M.; Habashy, T.M. Application of the multiplicative regularized contrast source inversion method on TM- and TE-polarized experimental Fresnel data. *Inverse Probl.* **2005**, *21*, S5–S13. [CrossRef]

27. Van den Berg, P.M.; van Broekhoven, A.L.; Abubakar, A. Extended contrast source inversion. *Inverse Probl.* **1999**, *15*, 1325–1344. [CrossRef]

28. Bloemenkamp, R.F.; Abubakar, A.; van den Berg, P.M. Inversion of experimental multi-frequency data using the contrast source inversion method. *Inverse Probl.* **2001**, *17*, 1611–1622. [CrossRef]

29. Jeffrey, I.; Zakaria, A.; LoVetri, J. Microwave Imaging by Mixed-Order Discontinuous Galerkin Contrast Source Inversion. In Proceedings of the 2014 XXXIst URSI General Assembly and Scientific Symposium (URSI GASS 2014), Beijing, China, 16–23 August 2014; pp. 255–258.

30. Lazebnik, M.; Popovic, D.; McCartney, L.; Watkins, C.B.; Lindstrom, M.J.; Harter, J.; Sewall, S.; Ogilvie, T.; Magliocc, M.A.; Breslin, T.M.; et al. A large-scale study of the ultrawideband microwave dielectric properties of normal, benign and malignant breast tissues obtained from cancer surgeries. *Phys. Med. Biol.* **2007**, *52*, 6093–6115. [CrossRef] [PubMed]

31. Kaye, C. Development and Calibration of Microwave Tomography Imaging Systems for Biomedical Applications Using Computational Electromagnetics. Master's Thesis, University of Manitoba, Winnipeg, MB, Canada, 2009.

32. Massey, F.J. The Kolmogorov-Smirnov Test for Goodness of Fit. *J. Am. Stat. Assoc.* **1951**, *46*, 68–78. [CrossRef]

33. Miller, L.H. Table of Percentage Points of Kolmogorov Statistics. *J. Am. Stat. Assoc.* **1956**, *51*, 111–121. [CrossRef]

34. Marsaglia, G.; Tsang, W.; Wang, J. Evaluating Kolmogorov's Distribution. *J. Stat. Softw.* **2003**, *8*, 1–4. [CrossRef]

35. MathWorks. Two-Sample Kolmogorov-Smirnov Test (R2018a). Available online: www.mathworks.com/help/stats/kstest2.html (accessed on 16 March 2018).

36. Kurrant, D.J.; Fear, E.C. Regional estimation of the dielectric properties of inhomogeneous objects using near-field reflection data. *Inverse Probl.* **2012**, *28*, 075001. [CrossRef]
37. Nemez, K.; Asefi, M.; Baran, A.; LoVetri, J. A faceted magnetic field probe resonant chamber for 3D breast MWI: A synthetic study. In Proceedings of the 2016 17th International Symposium on Antenna Technology and Applied Electromagnetics (ANTEM), Montreal, QC, Canada, 10–13 July 2016; pp. 322–324.

Journal of
Imaging

Article

Comparing Radar-Based Breast Imaging Algorithm Performance with Realistic Patient-Specific Permittivity Estimation

Declan O'Loughlin *, Bárbara L. Oliveira, Martin Glavin, Edward Jones and Martin O'Halloran

Electrical and Electronic Engineering, National University of Ireland Galway, Galway H91CF50, Ireland ;
barbara.oliveira@nuigalway.ie (B.L.O); martin.glavin@nuigalway.ie (M.G.); edward.jones@nuigalway.ie (E.J.);
martin.ohalloran@nuigalway.ie (M.O.)
* Correspondence: declan.oloughlin@nuigalway.ie

Received: 2 September 2019; Accepted: 11 November 2019; Published: 19 November 2019

Abstract: Radar-based breast imaging has shown promise as an imaging modality for early-stage cancer detection, and clinical investigations of two commercial imaging systems are ongoing. Many imaging algorithms have been proposed, which seek to improve the quality of the reconstructed microwave image to enhance the potential clinical decision. However, in many cases, the radar-based imaging algorithms have only been tested in limited numerical or experimental test cases or with simplifying assumptions such as using one estimate of permittivity for all patient test cases. In this work, the potential impact of patient-specific permittivity estimation on algorithm comparison is highlighted using representative experimental breast phantoms. In particular, the case studies presented help show that the permittivity estimate can impact the conclusions of the algorithm comparison. Overall, this work suggests that it is important that imaging algorithm comparisons use realistic test cases with and without breast abnormalities and with reconstruction permittivity estimation.

Keywords: radar-based breast imaging; microwave imaging; breast cancer

---

1. Introduction

Radar-based imaging is an emerging modality for breast cancer detection [1]. In recent years, a number of clinical studies have been published [2–7], including the commercially available MARIA® system (Micrima Ltd., Bristol, UK). A competing system developed by Microwave Vision Group (Villebon-sur-Yvette, France) is being used in clinical investigations at the National University of Ireland Galway (Ireland) [8–10]. The increasing amount of clinical studies of microwave breast imaging systems is motivating further research into the optimal imaging algorithms and system design which can maximize the clinical efficacy of radar-based breast imaging [11].

Several imaging reconstruction algorithms for radar-based imaging have been developed (known as beamformers) and many introductory books, comprehensive reviews and open-source implementations of these algorithms have been published [12–15]. These beamformers have also been compared using a variety of test cases with both numerical and experimental breast phantoms and patient images [16–22]. However, many comparative studies to date have used a limited set of beamformers [16], simplified numerical models [17,20], idealized artefact removal algorithms [18], or considered only test situations with abnormalities [21]. A more exhaustive comparative study using five clinical case studies identified that the Delay-Multiply-and-Sum (DMAS) algorithm achieved the highest signal-to-clutter ratios (SCRs) for the five clinical case studies, but did not consider the potential impact of variations in breast dielectric properties between the five patients [22].

However, it is known that an incorrect estimate of the breast dielectric properties can impair the image quality of radar-based breast images [23]. In particular, assuming the same estimate of breast dielectric properties for each test case has been shown to impair the sensitivity of radar-based breast imaging where tumors are not correctly detected [24]. Several methods to identify the optimal patient-specific dielectric properties have been proposed [25,26], many of which rely on the properties of the reconstructed image. Similarly, some preliminary studies have suggested that DMAS beamformer may "improve" the image quality in cases where no abnormalities exist [22,27]. However, to date, no study has analyzed the impact of different reconstruction algorithms on patient-specific permittivity estimation algorithms, nor has the potential benefit of improved imaging algorithms been examined with patient-specific estimation algorithms.

The goal of this study is to examine the impact of patient-specific permittivity estimation algorithms on beamformer comparisons in healthy and tumor experimental models. Although patient-specific permittivity estimation and sophisticated beamformers have been shown to improve image quality independently in certain cases, the impact of patient-specific permittivity estimation on beamformer comparative studies is unknown. The remainder of this paper is structured as follows: Section 2 reviews the beamformers which have been proposed in the literature as well as patient-specific permittivity estimation algorithms. Section 3 describes the experimental set-up used to test the algorithms. The results are presented in Section 4 and, finally, Section 5 concludes the paper.

2. Background

Many reviews, books and editorials have been published about microwave imaging and microwave breast imaging in recent years [1,11–15,28–37] which describe the mathematical background and existing literature of the method. Broadly speaking, microwave breast imaging algorithms attempt to infer information about the dielectric contrast of the breast, and correspondingly, information about any tumors or other abnormalities present. The subject of this work is radar-based imaging, a qualitative technique which reconstructs an energy map of the backscattered energy within the breast. As tumor tissues often have the highest dielectric properties [1], large reflections occur at tumor tissue boundaries. In areas of high dielectric contrast (i.e., tumors), these reflections cohere and result in high image energy, whereas in areas of lower contrast, these reflections do not cohere and result in lower image energy.

The Delay-and-Sum (DAS) imaging algorithm was the first proposed for radar-based breast imaging and has been used in both mono- and multistatic configurations [38,39]. Mathematically, DAS can be represented as follows:

$$I(\mathbf{r}) = w(\mathbf{r}) \int_{\Omega} P(\omega) \iint_{\mathcal{A},\mathcal{A}'} w_{\mathbf{a},\mathbf{a}'}(\mathbf{r}) E_{\mathbf{a},\omega}(\mathbf{a}') \exp\left[j\omega\tau_{\mathbf{a},\mathbf{a}'}(\mathbf{r},\omega)\right] \mathrm{d}\mathbf{a}\, \mathrm{d}\mathbf{a}'\, \mathrm{d}\omega \tag{1}$$

where the image intensity, $I(\mathbf{r})$, at the point $\mathbf{r}$ is calculated from the scattered energy, $E_{\mathbf{a},\omega}(\mathbf{a}')$, recorded at antenna $\mathbf{a}' \in \mathcal{A}'$ while the incident signal, $P(\omega)$, is transmitted from antenna $\mathbf{a} \in \mathcal{A}$ at angular frequency $\omega \in \Omega$. The propagation time for a signal from the transmitting antenna, $\mathbf{a}$, to the point of interest, $\mathbf{r}$, and back to the recording antenna, $\mathbf{a}'$, is represented by $\tau_{\mathbf{a},\mathbf{a}'}(\mathbf{r},\omega)$. DAS has been used in most patient imaging studies to date [3–5,40], and most comparative studies on imaging algorithm performance have used numerical or experimental data only [1,22].

Many extensions to DAS have been proposed which aim to improve the image quality. The extensions can be broadly classified into three types:

1. calculating a separate weighting factor to reward points where tumors are more likely based either on epidemiological studies or on characteristics of the scattered signals ($w(\mathbf{r})$ from Equation (1));
2. prioritizing scattered signals collected at certain locations based on the relative locations of the antennas and points of interest or on the antenna radiation patterns ($w_{\mathbf{a},\mathbf{a}'}(\mathbf{r})$ in Equation (1));

3. by improving the quality of the input signals prior to imaging through improved artefact removal algorithms or other noise reduction techniques.

In most cases, these modified beamformers have been tested in limited numerical or experimental test cases, and most radar-based operational systems have used traditional DAS [1]. In this work, the DAS beamformer is compared to the Delay-Multiply-and-Sum (DMAS) beamformer which has been identified as improving image quality in a series of clinical case studies [22]. In the literature, DMAS is increasingly being used in new experimental studies without comparison to DAS or evaluation in imaging scenarios without targets [41,42].

The DMAS beamformer was first proposed in [43]. DMAS aims to reduce noise in the scattered signals and hence, reduce clutter in the image. After synthetic focusing and prior to summation in Equation (1), the scattered signals are pairwise multiplied, such that:

$$I(\mathbf{r}) = \int_\Omega P(\omega) \iiiint_{A A' A A'} E_{\mathbf{a}_1,\omega}(\mathbf{a}_1') \exp\left[j\omega\tau_{\mathbf{a}_1,\mathbf{a}_1'}(\mathbf{r},\omega)\right] E_{\mathbf{a}_2,\omega}(\mathbf{a}_2') \exp\left[j\omega\tau_{\mathbf{a}_2,\mathbf{a}_2'}(\mathbf{r},\omega)\right] \, \mathrm{d}\mathbf{a}_1 \, \mathrm{d}\mathbf{a}_1' \, \mathrm{d}\mathbf{a}_2 \, \mathrm{d}\mathbf{a}_2' \, \mathrm{d}\omega \quad (2)$$

Although this does not increase the amount of independent information, this pairwise multiplication artificially increases the number of channels in the summation. The DMAS beamformer assumes that signals with a high degree of coherence should be rewarded by multiplication, whereas incoherent signals should not increase in energy after multiplication.

A fundamental assumption of radar-based imaging is that knowledge of propagation within the breast can be used to synthetically focus the recorded electromagnetic energy to points within the breast. This is represented by the propagation delay, $\tau_{\mathbf{a},\mathbf{a}'}(\mathbf{r},\omega)$, in Equation (1). The propagation delay can be calculated from the dielectric properties of the breast, however, in a realistic imaging scenario, the dielectric properties of the breast are not known. Thus, most practical implementations assume a single estimate of the dielectric properties for the entire breast, known as the effective permittivity in this work.

Several patient-specific permittivity estimation methods have been proposed which are reviewed in detail in [24]. Patient-specific permittivity estimation methods based on analyzing images reconstructed within a range of potential permittivity estimates have shown promising results in experimental and clinical test cases [23,26,44,45]. These patient-specific permittivity estimation methods have been shown to improve the sensitivity in 110 experimental test cases including diverse breast phantoms and tumor phantoms [46]. However, in [23,26], the original DAS beamformer was used, and the impact of other beamformers on the method is unknown.

Evaluation Metrics

To assess the performance of beamformers, it is necessary to identify a set of evaluation metrics for the radar-based images. In many cases, image quality has been defined using the signal-to-mean ratio (SMR) and the signal-to-clutter ratio (SCR). Typically, the SMR is calculated by dividing the mean energy of the tumor area (the area surrounding the maximum intensity of the image) by the mean energy of the image of the ipsilateral breast, where the tumor area is twice the full-width half-maximum (FWHM) of the response of highest energy in the image. Similarly, the SCR is calculated by dividing the maximum energy of the tumor area (i.e., the maximum intensity of the image) by the maximum energy in the background of the image. Additionally, in images of test scenarios with abnormalities, the localization error, or the distance between the maximum response in the image and the location of the abnormality, is used as a measure of quality. Lower localization error means higher image quality.

In this work, the SCR and the FWHM of the images using different beamformers is compared. Although recent studies have suggested that these metrics may not be sufficient to distinguish images of modelled healthy and cancerous cases [47], the SCR and the FWHM are still the standard metrics for image quality quantification.

Additionally, the images are analyzed quantitatively and annotated as containing a tumor or not according to the criteria described in detail in [24]. Broadly speaking, images where the main response has a SCR greater than 1.5 dB are annotated as a positive and considered to be detection. Where the main response is located within the physical extent of the tumor, the image is marked as a true positive, otherwise, it is considered a false positive.

3. Experimental Methods

In this section, the experimental methods are described. The BRIGID breast and tumor phantom set was used to assess the performance of the imaging algorithms which is fully presented in [48]. The breast and tumor phantoms are fabricated from a polyurethane mixture combined with graphite and carbon black powders to control the dielectric properties [49]. Once cured, the mixture is solid, and can be used to create a modular and stable test platform [48,50].

Breast phantoms were hemispherical with diameter of 14 cm and consisted of three tissue types:

- a hemispherical skin which varied between 1 to 3 mm in thickness with relative permittivity of $\varepsilon_r \approx 30$ and electrical conductivity of $\sigma \approx 2\,\mathrm{S\,m}^{-1}$ at 3 GHz;
- conical glandular structures to model breast lobes radiating from the areola with relative permittivity of $\varepsilon_r \approx 45$ and electrical conductivity of $\sigma \approx 2.5\,\mathrm{S\,m}^{-1}$ at 3 GHz;
- an adipose layer with relative permittivity of $\varepsilon_r \approx 8$ and electrical conductivity of $\sigma \approx 0.1\,\mathrm{S\,m}^{-1}$.

In each breast phantom, a hole was left in which a tumor "plug" could be inserted. Each tumor phantom was embedded in a plug of adipose tissue, and the tumor phantom had a mean relative permittivity of $\varepsilon_r \approx 70$ and mean electrical conductivity of $\sigma \approx 6\,\mathrm{S\,m}^{-1}$. The contrast between the dielectric properties of tumor and glandular tissues in these breast phantoms is similar to the contrast between healthy and glandular tissues in the literature [1,51–53].

The BRIGID phantom set contains breast phantoms which vary according to the volume of glandular tissue present within the phantom, known as the volume glandular fraction (VGF). Recent advances in three-dimensional imaging using breast tomosynthesis have allowed the VGF to be quantified, such as in a study of 219 women in [54]. This study suggests that breast density in terms of VGF may be overestimated from the two-dimensional slices used in X-ray mammography, and that the average VGF of breasts even in BI-RADS Class D maybe be as low as 20%.

In this work, images of four breast phantoms with VGF varying from 0% to 20% are analyzed as the healthy cases, specifically 0%, 10%, 15% and 20% as shown in Figure 1. These breast phantoms are representative of the variation in VGF present in the population in the sense that breast phantoms with VGF covering more than 80% of the population are included [54]. Each of the four breast phantoms can also be combined with a tumor phantom to model a cancerous case, the location of the tumor phantom is shown in Figure 1 by the dashed white circle. Five spherical tumor phantoms ranging in diameter from 5 mm to 20 mm are used, representing plugs 1 to 5 from [48], which are shown in Figure 1f. In total, there are 20 test cases with tumor phantoms (five tumor phantoms in four breast phantoms) and four test cases without tumors (four breast phantoms with an adipose tissue in place of the tumor phantom).

Signals are acquired from the breast phantoms using experimental hardware developed at NUI Galway, Ireland which is described fully in [23]. Experimental data between 2 GHz and 4 GHz were acquired using a ZNB40 2-port VNA and ZN-Z84 24-port switching matrix (Rohde & Schwartz GmbH, Munich, Germany) in the frequency domain. 24 flexible microstrip antennas were used to collect the multistatic scan. The antennas were first developed and used at McGill University in experimental and healthy clinical case studies [55,56]. The antennas housed in the radome are shown in Figure 1e. The antennas were designed to be in contact with a material with relative permittivity close to skin and were in direct contact with the skin of the phantom in this system. The imaging system was calibrated to the antenna using open/short/load and through measurements for all channels.

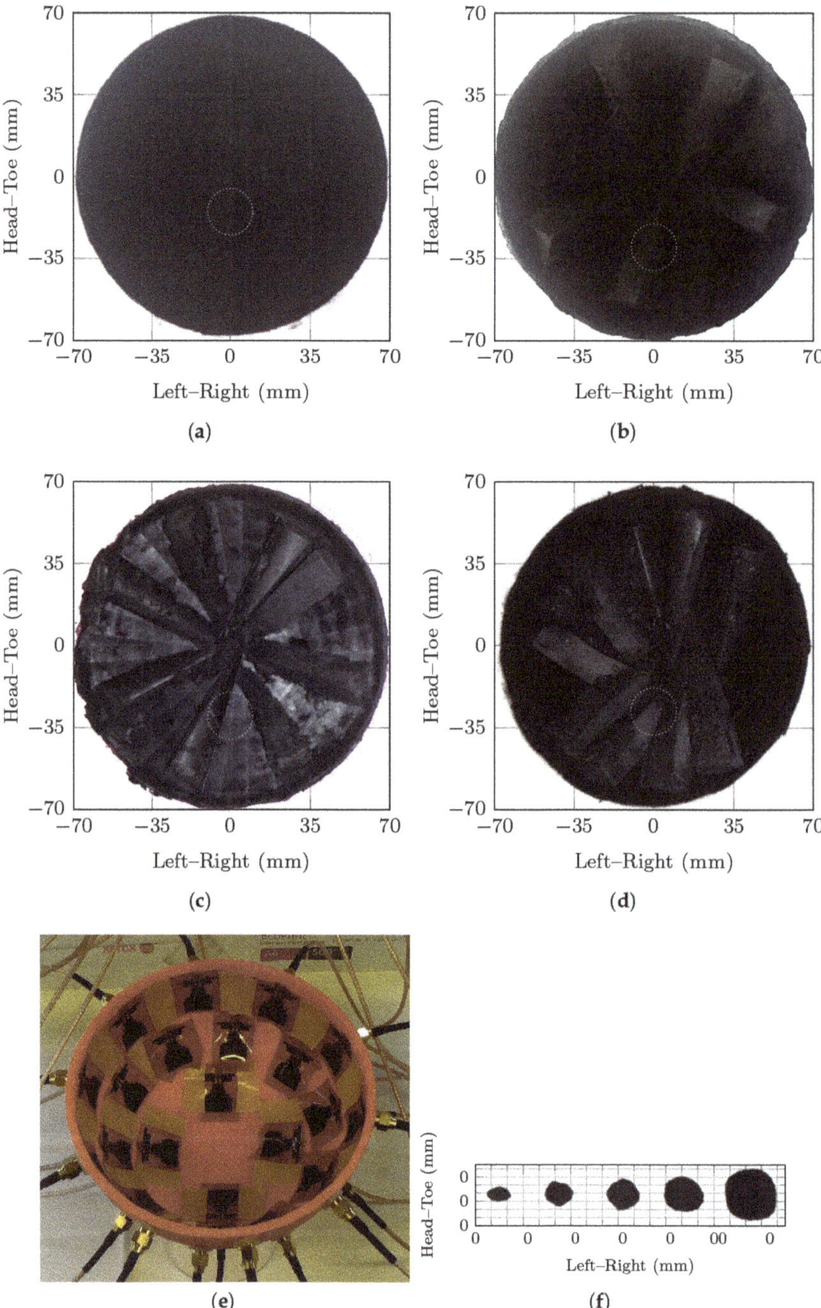

Figure 1. The four breast phantoms are shown in (**a–d**) containing 0%, 10%, 15% and 20% VGF, respectively. The tumor location is shown by the dashed, white circle. The 24 flexible microstrip antennas shown housed in the radome are shown in (**e**) and the 5 tumour phantoms are showin in (**f**).

A reference scan with a homogeneous breast phantom was used to account for differences between the antennas. This reference scan was used to calculate a "calibration factor" which was then applied to neighboring antennas such that the response of each antenna was the same. While effective in these experimental studies, this calibration is not optimized for use with patients without optimization to ensure the breast contacts the antennas correctly. Many approaches have been proposed in the literature to overcome these issues:

- inclusion of a suction system to ensure the breast contacts the radome [57];
- undersizing the radome compared to the reported breast size to improve the fit [56];
- use of a coupling medium to help reduce air gaps between the breast and the antennas [2,56,58];
- automated metrics to ensure the breast is well-situated prior to the scan [58].

After acquisition, all signal processing and imaging was performed in the frequency domain using MERIT, an open-source toolbox for microwave imaging [15]. First, rotational subtraction was performed to remove the large skin reflection and other artefacts [23]. In this experimental set-up, a rotational angle of 36 deg was used corresponding to the angle between neighboring antennas. Although rotational subtraction has also been used with the MARIA® system and hundreds of women [2,58], the optimal angle of rotation and the impact of rotational subtraction on image quality are not well studied.

Next, an assumed estimate of the relative permittivity of the breast was used to synthetically focus all signals to a point of interest within the breast [23]. Considering the generic radar-based beamformer shown in Equation (1), this is discretized and simplified to become the DAS beamformer in the frequency domain:

$$I(\mathbf{r}, \varepsilon'_r) = \sum_A \sum_{A'} E'_{\mathbf{a},\mathbf{a}'}(\mathbf{r}) \exp\left[\frac{j\omega}{c_0}\sqrt{\varepsilon'_r}\left(\|\mathbf{a}-\mathbf{r}\| + \|\mathbf{a}'-\mathbf{r}\|\right)\right] \tag{3}$$

where $I(\mathbf{r}, \varepsilon'_r)$ is the beamformer with the average permittivity of the breast, ε'_r, as a parameter; the antenna locations are the 24 antenna locations shown in Figure 1e; and the propagation paths are assumed to be the straight-line paths between the antennas and the points of interest. The DMAS beamformer is implemented similarly; however, prior to the beamformer being applied, the signals are pairwise multiplied.

For each test case, a cost function is used to estimate the "quality" of that image. Specifically, the Absolute Gradient, Φ^G_{DMA}, from [23] is applied. For each set of artefact-removed signals corresponding to a single test case, 50 images are reconstructed: specifically, both DAS and DMAS images are reconstructed at 25 permittivity values between $\varepsilon'_r = 8$ and $\varepsilon'_r = 14$. For each of the 24 test cases (5 tumor phantoms in 4 breast phantoms and the 4 breast phantoms without tumor phantoms), the 50 images per case are analyzed qualitatively and quantitavely below.

4. Results

The results of this paper are presented in a three stages:

1. first, the impact of permittivity estimation in phantoms without abnormalities is discussed in Section 4.1, considering both beamformers and the four breast phantoms;
2. secondly, the ability of DMAS to compensate for errors in the permittivity estimation process is considered in Section 4.2;
3. thirdly, the permittivity estimation algorithms are applied to images of the same scene reconstructed with different beamformers to estimate if the characteristics of the images are different in Section 4.3.

4.1. Algorithm Performance in Test Cases without Abnormalities

In this section, the impact of permittivity estimation on beamformer selection in breast phantoms without abnormalities is discussed. First, the images of the four breast phantoms without abnormalities using both the DAS and the DMAS beamformers are investigated in Figure 2. These images are reconstructed at $\varepsilon_r = 8.75$, which is the optimal permittivity based on maximizing the sensitivity for all test cases with abnormalities. Figure 2a,c,e,g are reconstructed with DAS, whereas Figure 2b,d,f,h are reconstructed with DMAS.

The images in Figure 2 highlight the potential challenges in distinguishing scenarios containing tumor phantoms from scenarios without tumor phantoms. Due to the reflections from other tissue boundaries within the breast (i.e., between the skin and the adipose tissue not fully removed by artefact removal and between the glandular structures and the adipose tissue), images of breast phantoms without tumor phantoms can have characteristics similar to images of tumor phantoms embedded in breast phantoms.

Considering the least dense breast phantom initially (Figure 2a,b), it can be seen that the DMAS image shows a distinct response in the middle of the breast whereas the DAS image of the same scene does not. A similar result is visible in Figure 2c,d where the DMAS image shows a more distinct response than the DAS image. These images suggest that DMAS may make the specificity of radar-based imaging worse even in breast phantoms with minimal dielectric contrast in the breast phantom interior.

However, in the denser breast phantoms, Figure 2e,f, both DAS and DMAS images show responses, as is expected as the fibroglandular structures have significant dielectric contrast to the adipose background. Interestingly, in the case of the 15% VGF breast phantom in Figure 2e,f, Where, as the DAS image highlights a response in the middle of the breast, the DMAS image highlights a response very close to the chest wall for the same scene. For the densest breast phantom in Figure 2g,h, it can be seen that the DMAS image shows a response in the same location as the DAS image, but with higher SCR.

Considering the images selected by using the permittivity estimation method described in [23], the parameter search algorithm selects $\varepsilon_r = 8$ for images of the 15% and 20% breast phantoms for both the DAS and DMAS images. In both cases, the responses are in the same location (within 5 mm) and could both be interpreted as containing a response that could be a tumor. However, the DMAS images have SCR of 2.62 dB and 3.06 dB for the 15% and 20% breast phantoms respectively, compared to 0.56 dB and 0.29 dB for the DAS images. These are shown quantitively in Table 1.

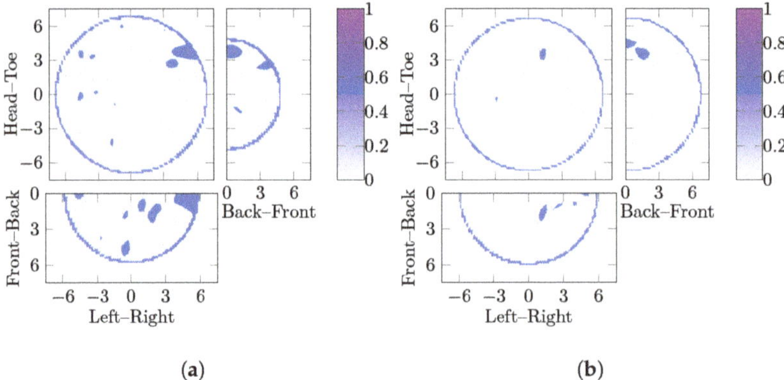

(a) (b)

Figure 2. *Cont.*

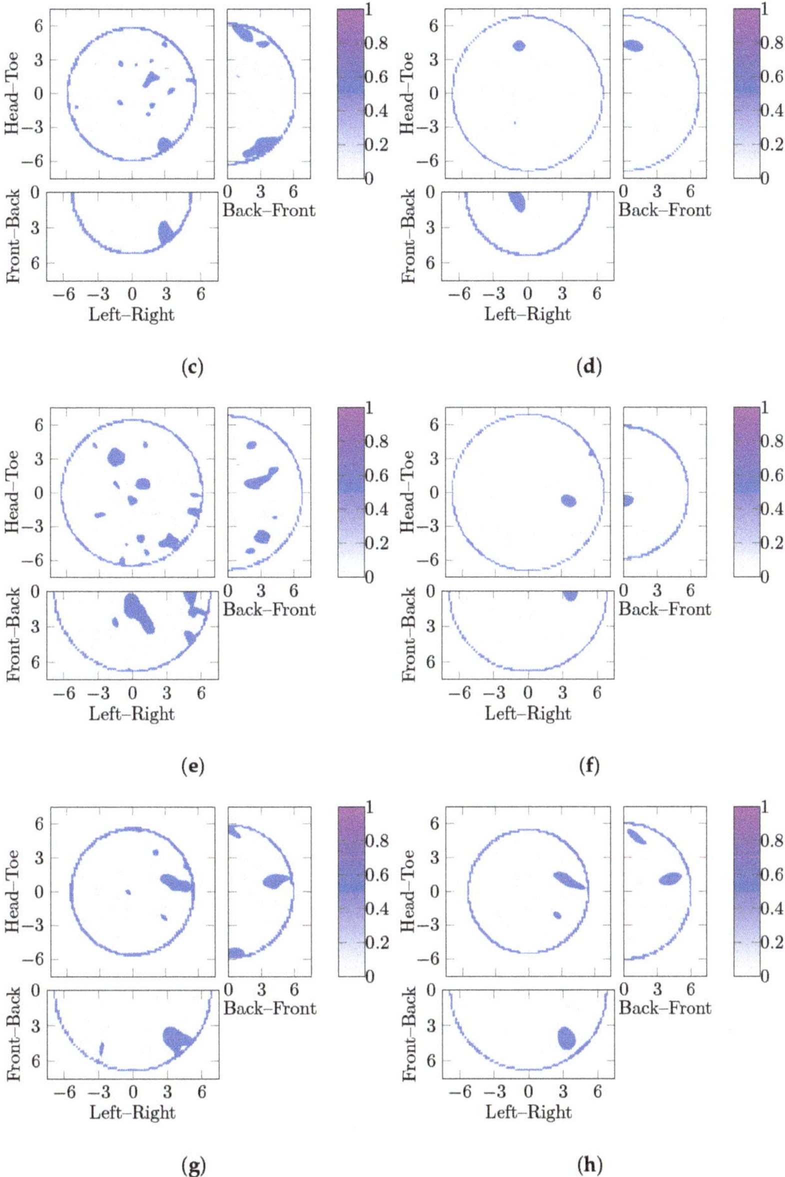

Figure 2. Images using DAS and DMAS of four breast phantoms with between 0% and 20% VGF. The images reconstructed using DAS are shown in the leftmost column (**a**, **c** and **e**) in order of increasing VGF (10%, 15% and 20% respectively). The corresponding image of the same test case reconstructed using DMAS is shown in the rightmost column (**b**, **d** and **f**). As a response that is very close to the skin would likely be regarded as an artefact, in the less dense breast phantoms, (a,c) reconstructed with DAS could be considered to be no detection, whereas the DMAS for the same breast phantom shows a response which is more in the center of the breast and more likely to be interpreted as a false positive. All dimensions are in cm, and the three slices shown (coronal, sagittal and axial) are cross-sections at the location of the maximum amplitude of the image in each case.

For the less dense breast phantoms, the parameter search algorithm would select images reconstructed at different permittivity values, $\varepsilon_r = 10.5$ and $\varepsilon_r = 9$ for DAS and DMAS respectively for the 0% breast phantom and $\varepsilon_r = 8.75$ and $\varepsilon_r = 8$ for the 10% breast phantom. No clear trend is visible: for 0% VGF the DAS image selected by the parameter search algorithm has higher SCR compared to the DMAS image selected by the parameter search algorithm (2.6 dB compared to 1.8 dB), whereas the reverse is true for 10% (1.6 dB compared to 2.0 dB). However, even in the simple and limited case studies shown in this section, the choice of beamformer may impact the specificity, suggesting that this should be evaluated in all beamformer comparison studies.

Table 1. The SCR and FWHM of images of breast phantoms without abnormalities containing 0% and 10% VGF, reconstructed with DAS and DMAS and a parameter search algorithm. For 0% VGF: $\varepsilon_r^{DAS} = 10.5$ and $\varepsilon_r^{DMAS} = 9$ whereas for 10% VGF, $\varepsilon_r^{DAS} = 8.75$ and $\varepsilon_r^{DMAS} = 8$.

	SCR (dB)				FWHM (mm)			
	0% VGF		10% VGF		0% VGF		10% VGF	
	ε_r^{DAS}	ε_r^{DMAS}	ε_r^{DAS}	ε_r^{DMAS}	ε_r^{DAS}	ε_r^{DMAS}	ε_r^{DAS}	ε_r^{DMAS}
DAS	2.6	1.5	1.6	0.75	10	12	13	7
DMAS	0.7	1.8	2.6	2.0	7	6	9	10

4.2. Effects of Permittivity Estimation on Performance

In this section, the effect of using DMAS instead of DAS are analyzed. The reconstruction permittivities are chosen using the parameter search algorithm applied to the DAS images. Additionally, case studies are presented which show how the beamformer comparison can depend on the permittivity chosen for reconstruction.

First, a case study is shown in Figure 3 where all six images are reconstructed from the same experimental case but using a reconstruction permittivity estimate of $\varepsilon_r = 11.75$ in Figure 3a,b, $\varepsilon_r = 12.25$ in Figure 3c,d, and $\varepsilon_r = 12.75$ in Figure 3e,f. Figure 3a,c,e are reconstructed with DAS, whereas Figure 3b,d,f are reconstructed with DMAS.

Considering the lowest estimate of $\varepsilon_r = 11.75$ in Figure 3a,b, the strongest response is correctly located in the images reconstructed with both DAS and DMAS. The FWHM is similar in both cases (approximately 8 mm), but the SCR of the image reconstructed with DMAS is 1.08 dB whereas the SCR of the image reconstructed with DAS is 0.09 dB.

Overestimating the relative permittivity by 5% in Figure 3c,d, the image reconstructed with DAS does not identify the tumor in the correct location (a false negative) whereas the image reconstructed with DMAS does correspond with the tumor location, albeit with a lower SCR of 0.42 dB. This result may indicate that improving the beamforming algorithm can compensate for errors in the relative permittivity estimates for image reconstruction.

Finally, overestimating the relative permittivity by 10% in Figure 3e,f, neither the image reconstructed with DAS nor the image reconstructed with DMAS show a response in the correct location. However, whereas the image reconstructed with DAS would be regarded as a false negative, the image reconstructed with DMAS does have a response in the image, but in the wrong location.

Considering all 20 cases (five tumor phantoms different in size in four breast phantoms differing in VGF), 14 are detected using DAS and permittivity estimation. For these 14, the DAS image is compared to the DMAS image at the same estimate of the permittivity. For 3 of the 14 images (P5 in 20%; and P3 and P4 in 15%), the DMAS image is "sharper" with SCR at least 1 dB higher and FWHM approximately 2 mm smaller. For a further 6 of the 14 (all from the 0% and 10% phantoms), the images were similar where the SCR was within 1 dB and the localization was within 1 cm. However, for 5 of the 14 cases (P1 from 0%; P1 from 10%; P2 and P5 from 15%; and P4 from 20%), the image deteriorated such that the tumor was not detected: for all five, the location of the maximum image response was more than 5 cm away from the location of the tumor phantom.

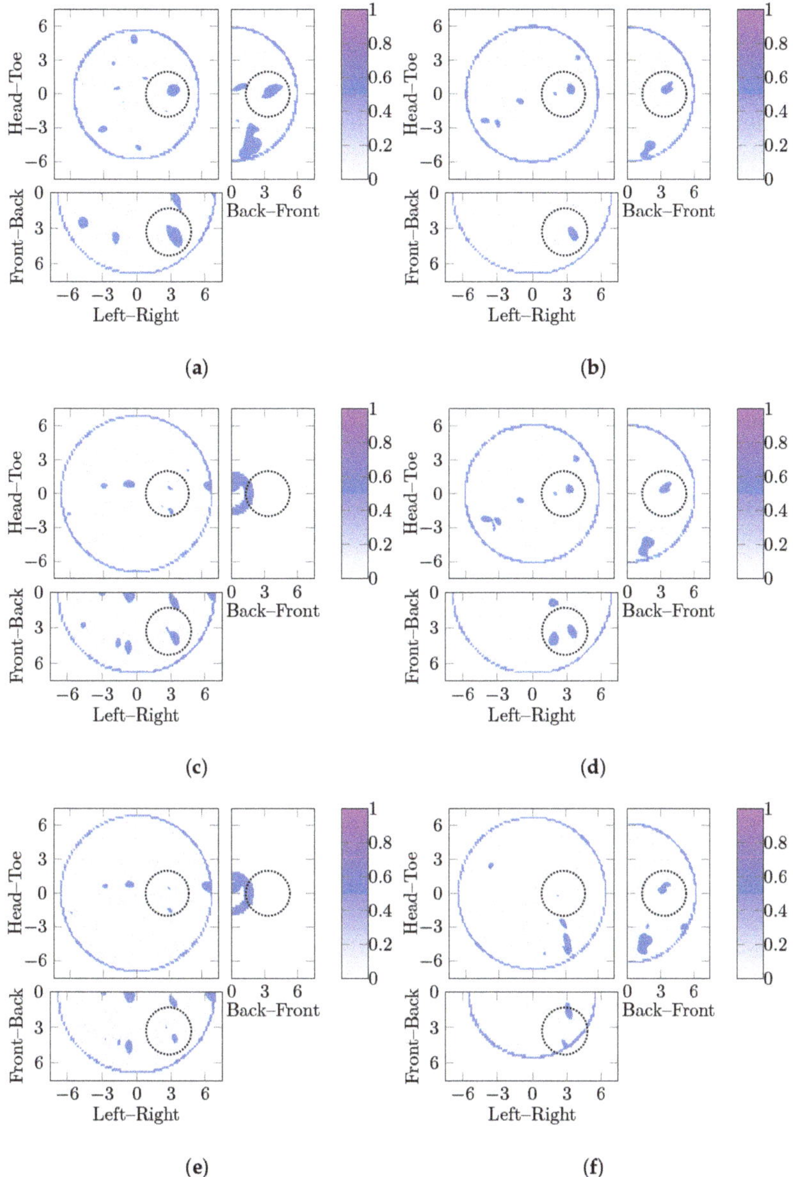

Figure 3. Six images of a 2 cm diameter tumor in the breast phantom with 20% VGF are shown, reconstructed with the DAS and DMAS beamformers are three different reconstruction permittivities. The DAS images are shown in the leftmost column in order of increasing permittivity, similarly, the corresponding DMAS images are shown in the rightmost column. Depending on the reconstruction permittivity chosen between $\varepsilon_i = 11.75$ and $\varepsilon_i = 12.75$, both beamformers perform the same and correctly localize the tumor in (**a**) and (**b**), DMAS outperforms DAS in (**c**) and (**d**), or neither DAS nor DMAS localize the tumor in (**e**) and (**f**). All dimensions are in cm, and the three slices shown (coronal, sagittal and axial) are cross-sections at the location of the maximum amplitude of the image in each case.

For the 6 of 20 cases where DAS and permittivity estimation did not detect the tumor, in 2 cases, the DMAS image was very similar to the DAS image but with less clutter. In 3 more cases, the DMAS image would also not have located the tumor in the right location, but the main response in the image was in a different location when compared to the DAS image. In only 1 case, P5 in 10% phantom, did the DMAS image reconstructed at the same permittivity as the DAS image correctly identify the tumor where the DAS image would not have.

4.3. Parameter Search Performance Using both Beamformers

Finally, in this section, images selected by DMAS and the permittivity estimation algorithm are compared to the DAS image at the same permittivity as the DMAS images, and to the DAS images selected by the permittivity estimation algorithm. Overall, the sensitivity from the DMAS images with permittivity estimation was 10 out of 20 cases compared to the 14 out of 20 for DAS with permittivity estimation.

For the 20 breast and tumor phantom combinations, in 9 cases the permittivity estimation algorithm selected a DMAS image reconstructed within $\Delta \varepsilon_r = 0.5$ of the DAS image. Of these 9 cases, 7 of the DAS images correctly identified the tumor in the correct location where as 6 of the DMAS images did. The majority of these nine cases were in the lower density phantoms: 6 of the 9 were images of phantoms with 0% and 10% VGF. This is expected, as images are less sensitive to the reconstruction permittivity at lower VGF.

For the remaining 11 cases where the reconstruction permittivity selected by the parameter search algorithm from the DMAS image differed from the DAS image, 7 of the 11 DAS images correctly identified the tumor whereas only 3 of the 11 DMAS images did. Interestingly, of the four images where DAS with permittivity estimation outperformed DMAS with permittivity estimation, two (P1 and P5) were in the breast phantom with 0% VGF.

5. Conclusions

In this work, the potential impact of permittivity estimation on beamformer comparative study design is highlighted. Although many comparative studies of various beamformers have been published in the literature, most use the simplifying assumption that the same estimate of the permittivity can be used for all test cases, although this is known to impair the image quality. Similarly, the majority include only test cases with tumors, and do not compare the beamformers in test cases without tumors.

The original DAS beamformer is compared to the DMAS beamformer, which was identified as a promising beamformer from a recent comparative study using five clinical case studies. The results in this paper suggest that DMAS may reconstruct an image of higher quality (defined primarily in terms of SCR) than the original DAS, however, this may result in more false positives such as in the less dense breast phantoms in this paper.

The importance of including realistic patient-specific estimation when comparing beamformers is also shown. A case study showing how the conclusion of a beamformer comparison study may vary depending on the permittivity estimate is presented. In this case study, as the estimate varies by 5%, it could be concluded that both beamformers perform the same, or that DMAS localizes the tumor where DAS does not, or that neither DAS nor DMAS localizes the tumor. Hence it is important to consider realistic and varied test cases when comparing beamformer performance.

Future work in this area should include the development of metrics to quantify the image quality which are useful to distinguish images of healthy and abnormal breasts. Future studies should consider the effects of permittivity estimation when designing and comparing beamformer performance experimentally and clinically.

Author Contributions: D.O. and B.L.O. designed and performed the experiments; D.O. analysed the data and wrote the paper; and M.G., E.J. and M.O. supervised the work, including experimental design, analysis and writing.

Funding: Irish Research Council (GOIPG/2014/987 and EPSPD/2019/200); Science Foundation Ireland (12/IP/1523); European Research Council (grant agreement number 637780).

Acknowledgments: This work was supported by the Irish Research Council (GOIPG/2014/987 and EPSPD/2019/200), Science Foundation Ireland (12/IP/1523) and the European Research Council (grant agreement number 637780).

Conflicts of Interest: The authors declare no conflict of interest.

References

1. O'Loughlin, D.; O'Halloran, M.; Moloney, B.M.; Glavin, M.; Jones, E.; Elahi, M.A. Microwave Breast Imaging: Clinical Advances and Remaining Challenges. *Trans. Biomed. Eng.* **2018**, *65*, 2580–2590. [CrossRef] [PubMed]
2. Preece, A.W.; Craddock, I.J.; Shere, M.; Jones, L.; Winton, H.L. MARIA M4: Clinical Evaluation of a Prototype Ultrawideband Radar Scanner for Breast Cancer Detection. *J. Med. Imaging* **2016**, *3*, 033502. [CrossRef] [PubMed]
3. Porter, E.; Coates, M.; Popović, M. An Early Clinical Study of Time-Domain Microwave Radar for Breast Health Monitoring. *IEEE Trans. Biomed. Eng.* **2016**, *63*, 530–539. [CrossRef] [PubMed]
4. Song, H.; Sasada, S.; Kadoya, T.; Okada, M.; Arihiro, K.; Xiao, X.; Kikkawa, T. Detectability of Breast Tumor by a Hand-Held Impulse-Radar Detector: Performance Evaluation and Pilot Clinical Study. *Sci. Rep.* **2017**, *7*. [CrossRef] [PubMed]
5. Yang, F.; Sun, L.; Hu, Z.; Wang, H.; Pan, D.; Wu, R.; Zhang, X.; Chen, Y.; Zhang, Q. A Large-Scale Clinical Trial of Radar-Based Microwave Breast Imaging for Asian Women: Phase I. In Proceedings of the International Symposium on Antennas and Propagation (APSURSI), San Diego, CA, USA, 9–14 July 2017; IEEE: San Diego, CA, USA, 2017; pp. 781–783. [CrossRef]
6. Song, H.; Sasada, S.; Masumoto, N.; Kadoya, T.; Shiroma, N.; Orita, M.; Arihiro, K.; Okada, M.; Kikkawa, T. Detectability of Breast Tumors in Excised Breast Tissues of Total Mastectomy by IR-UWB-Radar-Based Breast Cancer Imaging. *IEEE Trans. Biomed. Eng.* **2018**, *66*, 2296–2305. [CrossRef] [PubMed]
7. Wörtge, D.; Moll, J.; Krozer, V.; Bazrafshan, B.; Hübner, F.; Park, C.; Vogl, T. Comparison of X-ray Mammography and Planar UWB Microwave Imaging of the Breast: First Results from a Patient Study. *Diagnostics* **2018**, *8*, 54. [CrossRef]
8. Fasoula, A.; Anwar, S.; Toutain, Y.; Duchesne, L. Microwave Vision: From RF Safety to Medical Imaging. In Proceedings of the 11th European Conference on Antennas and Propagation (EuCAP), Paris, France, 19–24 March 2017; IEEE: Paris, France, 2017; pp. 2746–2750. [CrossRef]
9. Fasoula, A.; Duchesne, L.; Gil Cano, J.; Lawrence, P.; Robin, G.; Bernard, J.G. On-Site Validation of a Microwave Breast Imaging System, before First Patient Study. *Diagnostics* **2018**, *8*, 53. [CrossRef]
10. Duchesne, L.; Fasoula, A.; Kaverine, E.; Robin, G.; Bernard, J.G. Wavelia Microwave Breast Imaging: Identification and Mitigation of Possible Sources of Measurement Uncertainty. In Proceedings of the 13th European Conference on Antennas and Propagation (EuCAP), Kraków, Poland, 31 March–5 April 2019.
11. Bolomey, J.C. Crossed Viewpoints on Microwave-Based Imaging for Medical Diagnosis: From Genesis to Earliest Clinical Outcomes. In *The World of Applied Electromagnetics*; Lakhtakia, A., Furse, C.M., Eds.; Springer International Publishing: Cham, Switzerland, 2018; pp. 369–414. [CrossRef]
12. Nikolova, N.K. Microwave Biomedical Imaging. In *Wiley Encyclopedia of Electric and Electronics Engineering*; John Wiley & Sons, Inc: New York, NY, USA, 2014 ; pp. 1–22.
13. Conceição, R.C.; Mohr, J.J.; O'Halloran, M. (Eds.) *An Introduction to Microwave Imaging for Breast Cancer Detection*, 1st ed.; Biological and Medical Physics, Biomedical Engineering; Springer International Publishing: Cham, Switzerland, 2016.
14. Nikolova, N.K. (Ed.) *Introduction to Microwave Imaging*; EuMA High Frequency Technologies Series; Cambridge University Press: Cambridge, UK, 2017. [CrossRef]
15. O'Loughlin, D.; Elahi, M.A.; Porter, E.; Shahzad, A.; Oliveira, B.L.; Glavin, M.; Jones, E.; O'Halloran, M. Open Source Software for Microwave Radar Based Image Reconstruction. In Proceedings of the 12th European Conference on Antennas and Propagation (EuCAP), London, UK, 9–13 April 2018.
16. Klemm, M.; Craddock, I.J.; Leendertz, J.A.; Preece, A.W.; Benjamin, R. Radar-Based Breast Cancer Detection Using a Hemispherical Antenna Array—Experimental Results. *IEEE Trans. Antennas Propag.* **2009**, *57*, 1692–1704. [CrossRef]

17. Xie, Y.; Guo, B.; Xu, L.; Li, J.; Stoica, P. Multistatic Adaptive Microwave Imaging for Early Breast Cancer Detection. *IEEE Trans. Biomed. Eng.* **2006**, *53*, 1647–1657. [CrossRef]
18. O'Halloran, M.; Glavin, M.; Jones, E. Effects of Fibroglandular Tissue Distribution on Data-Independent Beamforming Algorithms. *Prog. Electromagn. Res.* **2009**, *97*, 141–158. [CrossRef]
19. Byrne, D.; O'Halloran, M.; Glavin, M.; Jones, E. Data Independent Radar Beamforming Algorithms for Breast Cancer Detection. *Prog. Electromagn. Res.* **2010**, *107*, 331–348. [CrossRef]
20. Moll, J.; Kexel, C.; Krozer, V. A Comparison of Beamforming Methods for Microwave Breast Cancer Detection in Homogeneous and Heterogeneous Tissue. In Proceedings of the Microwave Conference (EuMC), Nuremberg, Germany, 6–10 October 2013; pp. 1839–1842.
21. Elahi, M.A.; Lavoie, B.R.; Porter, E.; Glavin, M.; Jones, E.; Fear, E.C.; O'Halloran, M. Comparison of Radar-Based Microwave Imaging Algorithms Applied to Experimental Breast Phantoms. In Proceedings of the 32nd International Union of Radio Science (URSI) General Assembly and Scientific Symposium, Montreal, QC, Canada, 19–26 August 2017; Union of Radio Science (URSI): Montréal, QC, Canada, 2017.
22. Elahi, M.A.; O'Loughlin, D.; Lavoie, B.R.; Glavin, M.; Jones, E.; Fear, E.C.; O'Halloran, M. Evaluation of Image Reconstruction Algorithms for Confocal Microwave Imaging: Application to Patient Data. *Sensors* **2018**, *18*, 1678. [CrossRef] [PubMed]
23. O'Loughlin, D.; Oliveira, B.L.; Elahi, M.A.; Glavin, M.; Jones, E.; Popović, M.; O'Halloran, M. Parameter Search Algorithms for Microwave Radar-Based Breast Imaging: Focal Quality Metrics as Fitness Functions. *Sensors* **2017**, *17*, 2823. [CrossRef] [PubMed]
24. O'Loughlin, D.; Oliveira, B.L.; Santorelli, A.; Porter, E.; Glavin, M.; Jones, E.; Popović, M.; O'Halloran, M. Sensitivity and Specificity Estimation Using Patient-Specific Microwave Imaging in Diverse Experimental Breast Phantoms. *IEEE Trans. Med. Imaging* **2019**, *38*, 303–311. [CrossRef]
25. O'Loughlin, D.; Krewer, F.; Glavin, M.; Jones, E.; O'Halloran, M. Focal Quality Metrics for the Objective Evaluation of Confocal Microwave Images. *Int. J. Microw. Wirel. Technol.* **2017**, *9*, 1365–1372. [CrossRef]
26. Lavoie, B.R.; Okoniewski, M.; Fear, E.C. Estimating the Effective Permittivity for Reconstructing Accurate Microwave-Radar Images. *PLoS ONE* **2016**, *11*. [CrossRef]
27. O'Loughlin, D.; Glavin, M.; Jones, E.; O'Halloran, M. Evaluation of Experimental Microwave Radar-Based Images: Evaluation Criteria. In Proceedings of the Antennas and Propagation Society International Symposium (APSURSI), Boston, MA, USA, 8–13 July 2018; IEEE: Boston, MA, USA, 2018.
28. Iskander, M.F.; Durney, C.H. Electromagnetic Techniques for Medical Diagnosis: A Review. *Proc. IEEE* **1980**, *68*, 126–132. [CrossRef]
29. Larsen, L.; Jacobi, J. Microwaves Offer Promise as Imaging Modality. *Diagn. Imaging* **1982**, *11*, 44–47.
30. Fear, E.C.; Hagness, S.C.; Meaney, P.M.; Okoniewski, M.; Stuchly, M.A. Enhancing Breast Tumor Detection with Near-Field Imaging. *IEEE Microw. Mag.* **2002**, *3*, 48–56. [CrossRef]
31. Fear, E.C. Microwave Imaging of the Breast. *Technol. Cancer Res. Treat.* **2005**, *4*, 69–82. [CrossRef]
32. Nikolova, N.K. Microwave Imaging for Breast Cancer. *IEEE Microw. Mag.* **2011**, *12*, 78–94. [CrossRef]
33. Meaney, P.M. Microwave Imaging: Perception and Reality. *Expert Rev. Med. Devices* **2013**, *10*, 581–583. [CrossRef] [PubMed]
34. Crocco, L. Microwaves for Medical Imaging: Some Possible Pathways for an Accelerated Progress towards Clinical Practice. *New Horizons Transl. Med.* **2015**, *2*, 62. [CrossRef]
35. Chandra, R.; Zhou, H.; Balasingham, I.; Narayanan, R.M. On the Opportunities and Challenges in Microwave Medical Sensing and Imaging. *IEEE Trans. Biomed. Eng.* **2015**, *62*, 1667–1682. [CrossRef] [PubMed]
36. Kwon, S.; Lee, S. Recent Advances in Microwave Imaging for Breast Cancer Detection. *Int. J. Biomed. Imaging* **2016**, *2016*, 1–25. [CrossRef] [PubMed]
37. Modiri, A.; Goudreau, S.; Rahimi, A.; Kiasaleh, K. Review of Breast Screening: Towards Clinical Realization of Microwave Imaging. *Med. Phys.* **2017**, *44*, e446–e458. [CrossRef] [PubMed]
38. Li, X.; Hagness, S.C. A Confocal Microwave Imaging Algorithm for Breast Cancer Detection. *Microw. Wirel. Components Lett. IEEE* **2001**, *11*, 130–132.
39. Nilavalan, R.; Gbedemah, A.; Craddock, I.J.; Li, X.; Hagness, S.C. Numerical Investigation of Breast Tumour Detection Using Multi-Static Radar. *Electron. Lett.* **2003**, *39*, 1787–1789. [CrossRef]

40. Fear, E.C.; Bourqui, J.; Curtis, C.F.; Mew, D.; Docktor, B.; Romano, C. Microwave Breast Imaging With a Monostatic Radar-Based System: A Study of Application to Patients. *IEEE Trans. Microw. Theory Tech.* **2013**, *61*, 2119–2128. [CrossRef]
41. Shao, W.; Edalati, A.; McCollough, T.R.; McCollough, W.J. A Time-Domain Measurement System for UWB Microwave Imaging. *IEEE Trans. Microw. Theory Tech.* **2018**, *66*, 2265–2275. [CrossRef]
42. Islam, M.; Samsuzzaman, M.; Islam, M.; Kibria, S. Experimental Breast Phantom Imaging with Metamaterial-Inspired Nine-Antenna Sensor Array. *Sensors* **2018**, *18*, 4427. [CrossRef] [PubMed]
43. Lim, H.B.; Nhung, N.T.T.; Li, E.P.; Thang, N.D. Confocal Microwave Imaging for Breast Cancer Detection: Delay-Multiply-and-Sum Image Reconstruction Algorithm. *IEEE Trans. Biomed. Eng.* **2008**, *55*, 1697–1704. [CrossRef] [PubMed]
44. O'Loughlin, D.; Krewer, F.; Glavin, M.; Jones, E.; O'Halloran, M. Estimating Average Dielectric Properties for Microwave Breast Imaging Using Focal Quality Metrics. In Proceedings of the 10th European Conference on Antennas and Propagation (EuCAP), Davos, Switzerland, 10–15 April 2016; IEEE: Davos, Switzerland, 2016; pp. 1–5. [CrossRef]
45. O'Loughlin, D.; Glavin, M.; Jones, E.; O'Halloran, M. Optimisation of Confocal Microwave Breast Images Using Image Focal Metrics. In Proceedings of the 22nd Bioengineering in Ireland (BINI), Galway, Ireland, 22–23 January 2016; Royal Academy of Medicine in Ireland: Galway, Ireland, 2016; p. 39.
46. O'Loughlin, D.; Oliveira, B.L.; Glavin, M.; Jones, E.; O'Halloran, M. Advantages and Disadvantages of Parameter Search Algorithms for Permittivity Estimation for Microwave Breast Imaging. In Proceedings of the 13th European Conference on Antennas and Propagation (EuCAP), Krakow, Poland, 31 March–5 April 2019; IEEE: Kraków, Poland, 2019.
47. O'Loughlin, D.; Oliveira, B.L.; Glavin, M.; Jones, E.; O'Halloran, M. Effects of Interpatient Variance on Microwave Breast Images: Experimental Evaluation. In Proceedings of the 40th Annual International Conference of the Engineering in Medicine and Biology Society (EMBC), Honolulu, HI, USA, 18–21 July 2018; IEEE: Honolulu, HI, USA, 2018.
48. Oliveira, B.L.; O'Loughlin, D.; O'Halloran, M.; Porter, E.; Glavin, M.; Jones, E. Microwave Breast Imaging: Experimental Tumour Phantoms for the Evaluation of New Breast Cancer Diagnosis Systems. *Biomed. Phys. Eng. Express* **2018**, *4*, 025036. [CrossRef]
49. Garrett, J.; Fear, E.C. A New Breast Phantom With a Durable Skin Layer for Microwave Breast Imaging. *IEEE Trans. Antennas Propag.* **2015**, *63*, 1693–1700. [CrossRef]
50. Santorelli, A.; Laforest, O.; Porter, E.; Popović, M. Image Classification for a Time-Domain Microwave Radar System: Experiments with Stable Modular Breast Phantoms. In Proceedings of the 9th European Conference on Antennas and Propagation (EuCAP), Lisbon, Portugal, 13–17 April 2015; IEEE: Lisbon, Portugal, 2015; pp. 1–5.
51. Lazebnik, M.; McCartney, L.; Popović, D.; Watkins, C.B.; Lindstrom, M.J.; Harter, J.; Sewall, S.; Magliocco, A.; Booske, J.H.; Okoniewski, M.; et al. A Large-Scale Study of the Ultrawideband Microwave Dielectric Properties of Normal Breast Tissue Obtained from Reduction Surgeries. *Phys. Med. Biol.* **2007**, *52*, 2637–2656. [CrossRef]
52. Lazebnik, M.; Popović, D.; McCartney, L.; Watkins, C.B.; Lindstrom, M.J.; Harter, J.; Sewall, S.; Ogilvie, T.; Magliocco, A.; Breslin, T.M.; et al. A Large-Scale Study of the Ultrawideband Microwave Dielectric Properties of Normal, Benign and Malignant Breast Tissues Obtained from Cancer Surgeries. *Phys. Med. Biol.* **2007**, *52*, 6093–6115. [CrossRef]
53. Sugitani, T.; Kubota, S.; Kuroki, S.; Sogo, K.; Arihiro, K.; Okada, M.; Kadoya, T.; Hide, M.; Oda, M.; Kikkawa, T. Complex Permittivities of Breast Tumor Tissues Obtained from Cancer Surgeries. *Appl. Phys. Lett.* **2014**, *104*, 253702. [CrossRef]
54. Huang, S.Y.; Boone, J.M.; Yang, K.; Packard, N.J.; McKenney, S.E.; Prionas, N.D.; Lindfors, K.K.; Yaffe, M.J. The Characterization of Breast Anatomical Metrics Using Dedicated Breast CT. *Med. Phys.* **2011**, *38*, 2180–2191. [CrossRef]
55. Bahramiabarghouei, H.; Porter, E.; Santorelli, A.; Gosselin, B.; Popović, M.; Rusch, L.A. Flexible 16 Antenna Array for Microwave Breast Cancer Detection. *IEEE Trans. Biomed. Eng.* **2015**, *62*, 2516–2525. [CrossRef]
56. Porter, E.; Bahrami, H.; Santorelli, A.; Gosselin, B.; Rusch, L.A.; Popović, M. A Wearable Microwave Antenna Array for Time-Domain Breast Tumor Screening. *IEEE Trans. Med. Imaging* **2016**, *35*, 1501–1509. [CrossRef]

57. Kuwahara, Y. Microwave Imaging for Early Breast Cancer Detection. In *New Perspectives in Breast Imaging*; Malik, A.M., Ed.; InTechOpen: London, UK, 2017. [CrossRef]

58. Shere, M.; Lyburn, I.; Sidebottom, R.; Massey, H.; Gillett, C.; Jones, L. MARIA® M5: A Multicentre Clinical Study to Evaluate the Ability of the Micrima Radio-Wave Radar Breast Imaging System (MARIA®) to Detect Lesions in the Symptomatic Breast. *Eur. J. Radiol.* **2019**, *116*, 61–67. [CrossRef] [PubMed]

MDPI

St. Alban-Anlage 66

4052 Basel

Switzerland

Tel. +41 61 683 77 34

Fax +41 61 302 89 18

www.mdpi.com

Journal of Imaging Editorial Office

E-mail: jimaging@mdpi.com

www.mdpi.com/journal/jimaging